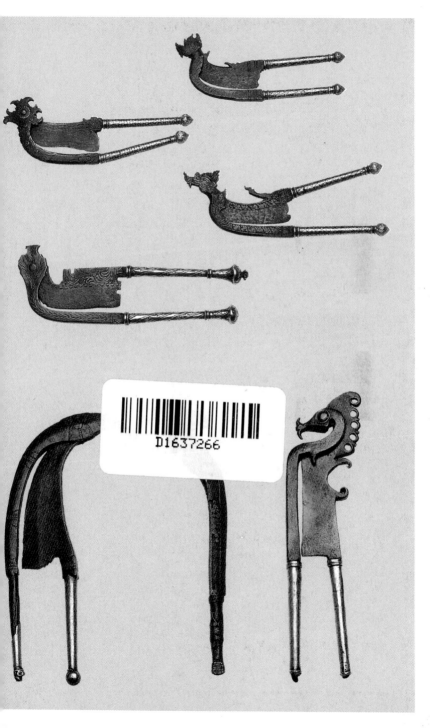

IMAGES OF ASIA

Betel Chewing Traditions
in South-East Asia

Titles in the series

Balinese Paintings (2nd edn.)
A. A. M. DJELANTIK

Bamboo and Rattan:
Traditional Uses and Beliefs
JACQUELINE M. PIPER

Betel Chewing Traditions
in South-East Asia
DAWN F. ROONEY

The Birds of Java and Bali
DEREK HOLMES and
STEPHEN NASH

The Birds of Singapore
CLIVE BRIFFETT and
SUTARI BIN SUPARI

The Birds of Sumatra and
Kalimantan
DEREK HOLMES and
STEPHEN NASH

Borobudur (2nd edn.)
JACQUES DUMARÇAY

Burmese Puppets
NOEL F. SINGER

Early Maps of South-East Asia
(2nd edn.)
R. T. FELL

Folk Pottery in South-East Asia
DAWN F. ROONEY

Fruits of South-East Asia: Facts
and Folklore
JACQUELINE M. PIPER

A Garden of Eden: Plant Life in
South-East Asia
WENDY VEEVERS-CARTER

Gardens and Parks of Singapore
VÉRONIQUE SANSON

The House in South-East Asia
JACQUES DUMARÇAY

Images of the Buddha in Thailand
DOROTHY H. FICKLE

Indonesian Batik: Processes,
Patterns and Places
SYLVIA FRASER-LU

Javanese Gamelan (2nd edn.)
JENNIFER LINDSAY

Javanese Shadow Puppets
WARD KEELER

The Kris: Mystic Weapon of the
Malay World (2nd edn.)
EDWARD FREY

Life in the Javanese Kraton
AART VAN BEEK

Mammals of South-East Asia
(2nd edn.)
EARL OF CRANBROOK

Musical Instruments of
South-East Asia
ERIC TAYLOR

Old Bangkok
MICHAEL SMITHIES

Old Jakarta
MAYA JAYAPAL

Old Malacca
SARNIA HAYES HOYT

Old Manila
RAMÓN MA. ZARAGOZA

Old Penang
SARNIA HAYES HOYT

Old Singapore
MAYA JAYAPAL

Sarawak Crafts: Methods,
Materials, and Motifs
HEIDI MUNAN

Silverware of South-East Asia
SYLVIA FRASER-LU

Songbirds in Singapore:
The Growth of a Pastime
LESLEY LAYTON

Betel Chewing Traditions
in South-East Asia

DAWN F. ROONEY

KUALA LUMPUR
OXFORD UNIVERSITY PRESS
OXFORD SINGAPORE NEW YORK
1993

Oxford University Press, Walton Street, Oxford OX2 6DP

Oxford New York Toronto
Delhi Bombay Calcutta Madras Karachi
Kuala Lumpur Singapore Hong Kong Tokyo
Nairobi Dar es Salaam Cape Town
Melbourne Auckland Madrid

and associated companies in
Berlin Ibadan

Oxford is a trade mark of Oxford University Press

Published in the United States
by Oxford University Press, New York

British Library Cataloguing in Publication Data

Data available

Library of Congress Cataloging in Publication Data

Rooney, Dawn.
Betel chewing traditions in South-East Asia/Dawn F. Rooney.
p. cm.—(Images of Asia)
Includes bibliographical references and index.
ISBN 0-19-588620-8 (boards):
1. Betel chewing—Asia, Southeastern—History. 2. Betel nut—Asia,
Southeastern—Folklore. 3. Decoration and ornament—Asia, Southeastern.
4. Asia, Southeastern—Social life and customs.
I. Title. II. Series.
GT3015.R66 1993
394. 1' 4—dc20
93–400
CIP

Typeset by Indah Photosetting Centre Sdn. Bhd., Malaysia
Printed by Kyodo Printing Co. (S) Pte. Ltd., Singapore
Published by Oxford University Press,
19–25, Jalan Kuchai Lama, 58200 Kuala Lumpur, Malaysia

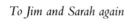

To Jim and Sarah again

Preface

EVER since I began writing this book, people have asked, 'How did you become interested in betel chewing?' While researching an earlier book (*Khmer Ceramics*, Oxford University Press, 1984), I discovered that one of the most common forms of glazed ceramics made by the ancient Khmers was a lime-pot. Traces of lime inside testify to its function as a container for lime paste, one of the main ingredients of a betel quid. Beyond that, I could find little information on the pots or their role in the practice of betel chewing. At the time, I had to continue with my writing on Khmer ceramics, but I vowed to return to the subject of betel chewing.

Through correspondence with over a hundred people connected with betel chewing in some way, my eyes were opened to the scope of the subject, most of which was as yet unexplored. Although considerable effort has been made to record the history of betel chewing on the Indian subcontinent where surviving artistic and literary sources are preserved in institutions and museums, in South-East Asia sources for studying the custom are limited and betel chewing seems to be decreasing more rapidly than in India and to a greater degree.

The main sources for this subject are the betel chewers of today. Since they are mostly elderly, the history of the custom threatens to die—unrecorded. This realization urged me to accelerate my research on the subject in the context of South-East Asia. My purpose in writing this book has thus been to bring the subject to the attention of those who are not familiar with it, to present the latest findings, and to bring together, for the first time, the art, legends, and lore of the custom. I hope it may stimulate further research into the history and traditions of betel chewing in South-East Asia.

Bangkok　　　　　　　　　　　　　　　DAWN F. ROONEY
July 1992

vii

Acknowledgements

I am grateful to the following for supplying photographs and for permission to reproduce them:

Cherie Aung-Khin (Elephant House, Bangkok), Colour Plates 10, 17, 19, and 22

Vance R. Childress, Colour Plates 2–9 and 13

Samuel Eilenberg and Hansjörg Mayer, Colour Plate 21, Plate 20, and Endpapers

Sylvia Fraser-Lu, Colour Plate 18 and Plate 23

Industrial Finance Corporation of Thailand, Colour Plate 12

Mohd. Yunus Noor, Colour Plate 14 and Plate 21

Donald L. Petrie and Niyanee Srikanthimarak, Colour Plates 23 and 24 and Plates 14, 25, 26, and 29

Rijksmuseum-Stitching, Amsterdam, Plate 27

River City Auction, Colour Plate 26

Suan Pakkad Palace, Colour Plate 25

Michael J. Sweet, Plates 1–5, 7–11, and 28

Widhaya Chaicharnthipayudh, Plate 24

Yadasji Wongpaiboon, Plate 22

Yanyong Manoopol, Plate 6

And to the following for permission to reproduce excerpts from material in copyright:

Ohio University Press

School of Oriental and African Studies, University of London

I would also like to express my appreciation to the following who have given me advice and support in my research:

Terese Bartholomew, Asian Art Museum of San Francisco

Angus Forsyth

John Guy, Victoria & Albert Museum

Dr Angela Hobart

Jyotindra Jain, National Handicrafts & Handlooms Museum

Janine Schotsmans, Musées Royaux d'Art et d'Histoire

Michael J. Sweet
Dr Andrew Turton, School of Oriental and African Studies
John van der Linden
Michael Wright

Contents

	Preface	*vii*
	Acknowledgements	*ix*
1	The Tradition	*1*
2	The Quid	*16*
3	The Symbolism	*30*
4	The Art	*40*
5	Conclusion	*66*
	Appendix	*68*
	Select Bibliography	*69*
	Index	*74*

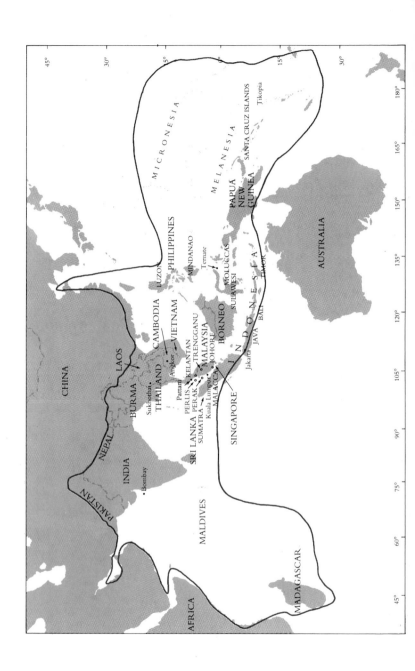

1
The Tradition

FEW traditions in South-East Asia have the antiquity and universal acceptance of betel chewing. The custom is over 2,000 years old and has survived from ancient times into the twentieth century. Its use cuts across class, sex, or age: 'The habitual and universal solace of both sexes is the areca nut and betel . . . which is rarely absent from the mouth of man or woman,' wrote the Honourable George N. Curzon, a nineteenth-century observer. Its devotees include farmers, priests, and kings, men, women, and children. The homeliness of the name belies its importance. The Thais 'prefer to go without rice or other food rather than to deprive themselves of the betel . . .' noted Nicolas Gervaise, a French visitor in the seventeenth century.

Three ingredients—an areca-nut, a leaf of the betel-pepper, and lime—are essential for betel chewing; others may be added depending on availability and preference. The leaf is first daubed with lime paste and topped with thin slices of the nut, then it is folded or rolled into a bite-size quid. The interaction of the ingredients during chewing produces a red-coloured saliva. 'If a person speaks to you while he is chewing his "quid" of betel, his mouth looks as if it were full of blood,' reflected Isabella L. Bird, an intrepid woman traveller of the nineteenth century. Most of the betel juice is spat out. The tell-tale residue looks like splotches of dried blood. Indeed, the resemblance is so close that some early European visitors thought many Asians had tuberculosis. The splotches of betel spittle are spaced consistently enough for use as measurements of time and distance in rural areas. A short time is 'about a betel chew' and the distance between two villages, for example, may be 'about three chews'.

Early Europeans called the custom 'betel-nut chewing'. The term, though, is incorrect because an areca-nut, not a betel-nut is chewed. Although 'betel-nut' continued as an entry in many

English language dictionaries until recently, nowadays the custom is defined correctly under 'betel'. This book thus uses the terms 'betel' and 'betel chewing', except in quotations or captions where the original text is retained.

Besides being chewed, the betel quid and the individual ingredients are widely used for medicinal, magical, and symbolical purposes. It is administered as a curative for a plethora of ills, including indigestion and worms. It is believed to facilitate contact with supernatural forces and is often used to exorcize spirits, particularly those associated with illness. In its symbolical role, it is present at nearly all religious ceremonies and festivals of the lunar calendar. Betel fosters relationships and thus serves as an avenue of communication between relatives, lovers, friends, and strangers. It figures in male–female alliances and its potency in this area is especially telling. Because of its power in bonding relationships, betel is used symbolically to solidify acts of justice such as oaths of allegiance and the settlement of lawsuits. Betel is a surrogate for money in payment to midwives and surgeons for services rendered.

A key to the unconditional patronage of betel is its use on four levels—as a food and medicine, and for magical and symbolical purposes. As such, this single tradition is an integral part of the art, ceremonies, and social intercourse of daily life.

Geographically, the use of betel is widespread: its parameters encompass the eastern coastline of Africa to Madagascar in the West; Melanesia to Tikopia (in the Santa Cruz Islands) in the East; southern China in the North, and Papua New Guinea in the South. This area includes the Indian subcontinent, Sri Lanka, and all of South-East Asia. The boundaries extend from longitude 170° E to 40° W, and from latitude 40° N to 15° S (see map). Within this area, betel chewers comprise over one-tenth of the world's population.

The most concentrated areas for betel chewing are where the climate and soil are suitable for the cultivation of the nut and the leaf, and where there is an adequate source of lime. Although betel chewing is common in the coastal districts of southern China, for example, it is rare further inland because of the difficulty in obtaining the ingredients. Likewise, betel chewing is less common in the interior of Sumatra because of the lack of a source of lime. Some areas included in the betel chewing nucleus do not produce the ingredients but can obtain them through internal or regional trade.

The tropical climate of South-East Asia fosters the cultivation of the nut and the leaf. A grove of areca-nut palms (*Areca catechu* Linn.) stands majestically in an idyllic island setting of an early nineteenth-century drawing (Plate 1, on the right). Travelling on the river from the mouth of the Gulf of Thailand to the former capital of Ayutthaya in the seventeenth century, Gervaise noted 'an oases of dark-green areca ... marks the site of villages ...' and on the river approach to Bangkok, the boat of Maxwell Sommerville 'passed large plains of betel-nut palms ...'. Over-looking the city of Bangkok, some hundred years ago, 'a waving sea of cocoa-nut and betel-nut palms is about all that distinctly appears', observed Frank Vincent. Travelling by elephant from Bangkok to Nakorn Ratchasima (Korat) in central Thailand in the last century, James McCarthy, an Irishman who was Director-General of the Royal Survey Department of Siam, saw 'the areca palms gracefully swaying to and fro over houses that nestle in the shade ...'.

A betel set is a necessary accompaniment for chewing a quid. It serves as a container for storing the ingredients and ensures they are always in one place and ready for use. A basic set includes a tray, individual containers, and a tool for cutting the nut. The material and workmanship of a set vary, ranging from simple to complex. Those of the agrarian population reflect an unsophist-icated art distinguished by an honesty of purpose. Elaborate betel sets owned by royalty are made of the finest materials available, often gold or silver and inlaid with precious stones. Both types survive as legacies of the betel chewing tradition in South-East Asia and give us an insight into the social and cultural milieu of the people who used them.

Betel chewing, so firmly embedded in the traditions of South-East Asia, was a custom totally alien to early European witnesses. Their impressions were unfavourable, yet they were all impelled

1. 'Cocoa Nut & Betel Trees', from Thomas and William Daniell, *A Picturesque Voyage to India by way of China*, London, 1810. (Courtesy Antiques of the Orient, Singapore)

to write about it. Betel chewing is an 'unhygienic, ugly, vile, and disgusting' habit wrote Mary Cort. 'This abominable practice of betel nut chewing.... It is a revolting habit,' endorsed Sommerville. In light of these views, it is surprising that some Europeans did adopt the habit. Portuguese women living in India '... have the like custome of eating these Bettele leaves, so that if they were but one day without eating their Bettele, they persuade themselves they could not live ...', and Portuguese men '... by the common custome of their wives eating of Bettele, doe like-wise use it', wrote John Huyghen van Linschoten, a Dutch traveller to the East in the late sixteenth century.

Why do people chew betel? The multi-purpose benefits are described explicitly in Indian literature as early as the sixth century. 'Betel stimulates passion, brings out the physical charm, conduces to good-luck, lends aroma to the mouth, strengthens the body and dispels diseases arising from the phlegm. It also bestows many other benefits.' According to a sixth-century Indian text (quoted in Morarjee, n.d.), betel is one of the eight enjoyments of life—along with unguents, incense, women, garments, music, beds, food, and flowers named in a Sanskrit verse of the twelfth century. Chou Ta-Kuan (Zhou Daguan), a member of a Chinese mission to Angkor at the end of the thirteenth century, gives more practical reasons for chewing betel: he thought it prevented belching after meals. Betel is 'energy giving medicine' for Khun (Mrs) Samap, an 85-year-old Thai barber, who always chews it before giving a haircut. It 'encourages self-reflection' for a Burmese monk.

The main reason for chewing betel seems to lie in the social affability produced by sharing a quid with friends. This enjoyment can be seen on the faces of a group of elderly men squatting around a betel box, or heard in the laughter of women relaxing in a rice-field with a betel basket. Offering a quid to someone is a mark of hospitality. 'When they receive passing guests, they entertain them, not with tea, but only with areca nut,' observed the Chinese Ma Huan on a naval expedition to the 'Southern Ocean' in the fifteenth century. Afterwards, 'O friend, there are hundreds of thousands [of] good qualities in tambula [betel]; there is, however, one immensely bad feature associated with it, namely, the sending away [of friends] after its bestowal,' extols a Sanskrit verse (Morarjee, n.d.).

European accounts give reasons for chewing betel ranging from the ridiculous to the insightful. Antonio Pigafetta, an Italian member of the Ferdinand Magellan expedition to the Philippines in 1521, for example, thought people used betel '... to fortify their heart. If they abstained from it, they would die.' An equally absurd idea was put forth by P. A. Thompson that 'the quid is stuffed away in the cheek, and only so can the pure Siamese accent be produced....' Johan Nieuhof wrote in the seventeenth century that betel chewing '... corrects a stinking breath, cures the tooth-ache and scurvy, fastens the teeth and strengthens the gums ...'. This is one of many mentions in Western accounts of the benefits of betel as a breath freshener. 'Asian women would not think of making love without first sweetening their breath with betel, and the Portuguese and Dutch women ... quickly adopted the same practice,' wrote E. de Haan.

Some Europeans wrote erroneously of the relationship, or lack of, between betel chewing and kissing. According to Curzon, 'It is to the same practice [betel chewing] that I suppose must be attributed the total absence ... of that agreeable mark of salutation ... the kiss. Lips so tainted could hardly embrace.' Cort had a similar impression: 'The imported kiss is not yet in vogue and I do not see that it ever can be until betel is discarded, for at the present, the nose is a more kissable feature of the Siamese face than the mouth.' The 'absence' of the kiss is more likely due to the social attitude of the Thais rather than to the tradition of betel chewing. The traditional form of greeting for the Thais has always been a *wai* executed with the hands held in a prayer-like position, rather than a kiss or a handshake.

Red lips are a desirable mark of beauty in South-East Asia, just as in most parts of the world. A Burmese love song sings of 'lips reddened by betel juice ...'. Until the advent of lipstick, this look was achieved by chewing betel regularly. Young Burmese girls reddened their lips with betel for 'dressing up' occasions, recalls Khin Myo Chit. An Akha hill tribe woman, living amongst the ethnic minorities in the mountainous areas of northern Thailand, proudly displays her betel-reddened lips (Colour Plate 1). Asian women admired and envied the 'red lips' of European ladies. 'The Vermilion Lips, which the Siamese saw in the Pictures of our Ladies which we had carried to this country, made them to say that we must needs have in France, better Betel than theirs,' wrote Simon de la Loubère, envoy of the King of France in the seventeenth century.

The folklore and literature of the region reflect the venerated status of betel in the culture of South-East Asia. A favourite children's story recalls a young boy walking in the forest looking for wood. He comes upon a nest of parakeets. 'Now, what a fortune,' he thinks. 'If I get all these birds I can sell them for a good price and then try to buy myself cloth and betel and areca nut.'[1] Another legend describes a duel between Hang Tuah, a famous warrior, and Hang Jebat at the palace of the Sultan of Malacca. During the duel, Hang Tuah calls for 'time-out' to chew a quid of betel. Hang Jebat watches him walk to the sidelines. His eyes return to his opponent's unguarded magic kris. Knowing its power, he rushes up and steals the kris. Hang Jebat wins the duel, not by strength or skill but because Hang Tuah stopped to chew betel.

Betel Chewing and Royalty

Betel chewing 'prevailed especially among the nobles and magnates and kings', observed Marco Polo in the thirteenth century (Latham, 1958). In the same century, the King of Pagan decreed that anyone using the gilded pillars in the halls of temples for cleaning his fingers after chewing betel would be punished by having the index finger of his right hand cut off. Forgetting his royal edict, the King rubbed his betel-stained fingers on the gilded posts, and when reminded of the decree endured the punishment of cutting off his own finger.

[1]A. Soebiantoro and M. Ratnatunga, *Folk Tales of Indonesia*, New Delhi, Sterling, 1977.

From an equally early date, betel was used as a social denominator and a symbolical element for solidifying relationships amongst royalty in the region. A. Teeuw and D. K. Wyatt tell in *The Story of Patani* how the queen welcomed the Sultan of Johore by bringing him 'all kinds of food as well as betel leaf with areca nuts'. In the fourteenth century, the King of Pegu met the leader of the Shan States and they 'exchanged their betel boxes, spittoons and such articles of pomp, and delimited the frontier together'.

From the sixteenth century onwards, when Europeans reached the East, accounts are rich with descriptions of the royal use of betel. Every visitor had an audience with the king upon arrival and the presence of betel was a social custom. Henri Mouhot, the French naturalist attributed with the rediscovery of Angkor, wrote in his diary of 1862: 'Yesterday I was presented to the King [of Luang Prabang, Laos], who received me with a great display of pomp and splendour; he was surrounded by mandarins and ill-clad guards. His Majesty sat upon a kind of sofa-throne, chewing the betel-nut, and making all sorts of grimaces.' Plate 2 shows the King of Siam seated on his throne; the royal betel set is to his right and a gold spittoon is near his feet.

The custom itself was appalling enough for Europeans, but one aspect must have been nearly intolerable. Van Linschoten wrote that there was no greater honour for a Westerner than if the king 'profereth him of the same Bettele that he himselfe doth eate'. Alexander Hamilton described, in 1727, how this was done: 'In [a] visit with the King [of Quedah] ... he honours the guest with a seat near him and will chew a little betel and spit it out on [a] little gold saucer and sends it by page to the guest who must take it with all signs of humility and satisfaction and chew it after him; very dangerous to refuse royal morsel.'

The king was always attended by 'betel slaves', whether he was in the palace or travelling in the area. Plate 3, a print by Theodore de Bry in *Petits Voyages*, shows a 'betel slave' handing leaves to a ruler on an outing. 'Noblemen and Kings, wheresoever they goe, stand or sit, have alwaies a servant by them, with a Silver ketle full of Bettele and their mixtures, and give them a

2. 'Supreme King of Siam in His State Robes', from Frank Vincent, *The Land of the White Elephant: Sights and Scenes in South-East Asia 1871–1872*, New York, Harper and Brothers, 1874. (Courtesy Antiques of the Orient, Singapore)

3. 'Betel slaves' handing betel leaves to an Asian ruler, from *Petit Voyages* (Latin edition), illustrated by Theodore de Bry, Frankfurt am Main, 1601. (Courtesy Antiques of the Orient, Singapore)

leafe ready prepared,' wrote van Linschoten. Three hundred years later, Sommerville witnessed a similar scene: 'Men of rank and opulence are always accompanied by a servant, who carries his master's areca or betel-nut box.' Major F. McNair observed, in 1878, that 'when the Rajah of Perak in Malaysia chews betel his wives are stationed behind him beating up the components to save trouble in mastication'. Antonio Galvao reported in the sixteenth century that the King of Ternate allegedly engaged female dwarfs that were crippled in childhood to carry his set.

Betel sets were exchanged as gifts between foreign rulers. When Justus Schouten, manager of the Dutch East India Company at Ayutthaya, left for Batavia (now Jakarta) in 1634, the King of Thailand allowed him to carry the 'Royal Silver betel-box'. A gold betel box presented by the Sultan of Perak from Malaysia to the Prince and Princess of Wales when they visited Singapore in 1901 is illustrated in Plate 4.

4. 'Golden betel box' gift from the Sultan of Perak to the Prince and Princess of Wales in 1901, from H. Ling Roth, *Oriental Silverwork: Malay and Chinese*, London, Truslove & Hanson, Ltd., 1910. (Courtesy Antiques of the Orient, Singapore)

A betel set was an 'indispensable insignia of office, or of social rank', noted Carl A. Bock, a Norwegian natural scientist. A person's position could be identified by the material of his betel set and the degree of decoration. Members of a mission from Thailand in the seventeenth century prostrate at an audience with a monarch of India in Plate 5. A betel box of fine metal on a pedestal stands beside each member and signifies his position. The Sultan of Malacca regularly rewarded his ministers for loyal and distinguished services with the presentation of a betel set. Regulations set out by the second ruler of Malacca in the middle of the fifteenth century listed the names of high officials to whom betel quids could be given, and the order in which these were to be distributed. Betel sets in Burma were part of the regalia by the fifteenth century and shapes conformed to a hierarchical order. Sets with eight or twelve divisions usually signify a senior prince

11

5. Members of a mission from Thailand, each with a betel set, prostrate
before a monarch of India, from John Harris, *A Complete Collection of
Voyages and Travels*, London, 1764. (Courtesy Antiques of the Orient,
Singapore)

in Burma, whereas a betel set designated for the Heir Apparent or
a Queen is in the shape of a palace.

Betel sets gradually became part of the regalia for royalty in
South-East Asia, and remained so into the twentieth century. At a
ceremony installing the ten-year-old Crown Prince of Thailand
in 1878, he 'was carried on a gold chair, preceded by five girls,
dressed like angels, bearing his gold betel box, tea pot and
other utensils ...' (R. Brus, 1985). A betel set with enamelled
containers inlaid with rubies and a gold tray was part of the royal
regalia when the present Crown Prince of Thailand was installed.
The Raja of Perlis in Malaysia uses an elaborate gold betel set of

historical significance for court ceremonies. The set was presented to Malaysia by the King of Thailand when he was on a tribute mission in 1842. An ornamental gold betel box is among the seven articles that must be carried at an installation ceremony for the Sultan of Trengganu.

Origins

The origins of betel chewing are unknown. Research into the source is complicated by the fact that three ingredients are used. A further difficulty lies in determining whether seeds of the nut are indigenous or were transported to the place of discovery. Although it has long been held that betel chewing is native to India, recent linguistic and archaeological evidence casts doubt on this theory. Only literary evidence continues to support an Indian origin.

The word 'betel' was first used in the sixteenth century by the Portuguese. According to I. H. Burkill, it is probably a transliteration of the Malay word *vetila* ('the mere leaf') which is close in sound to 'betel'. Since its earliest use, the word has undergone a series of spellings from 'bettele' to 'betre' to 'betle' and finally to 'betel'. 'Areca' may have derived from the Malay word *adakka* ('areca-nut') or from *adakeya*, the Indian equivalent.

The widest range of words for 'areca' and 'betel' has been found in Indonesia, which suggests it may be the original location where these words were spoken. In India, on the other hand, the lack of variety of words for 'areca' and 'betel' indicates a later date of origin for the plants in that area. (For a list of regional words used today, see the Appendix.) Moreover, *sireh*, the most widespread name for 'betel' in Malaysia, is not derived from Sanskrit, which, according to N. M. Penzer, suggests betel chewing might have developed independently in Malaysia. Based on linguistic evidence, therefore, the custom seems to be native to the Indonesian archipelago.

The earliest archaeological evidence found so far is at Spirit Cave in north-western Thailand, where remains of *Areca catechu* dating from 10,000 BC have been found (Gorman, 1970). Similar remains have been found at other early sites in Thailand, including Ban Chiang, dated from 3600 BC to AD 200–300 (White, 1982). All finds, however, are from the cultivated plant; the absence of a wild species in the same area may suggest the custom originated elsewhere. The wild species has been found in Malaysia and adds archaeological support to the linguistic evidence of its origin in that area. Skeletons bearing evidence of betel chewing, dated to about 3000 BC, have also been found in the Duyong Cave in the Philippines (Bellwood, 1979). Compared with these finds, the earliest archaeological evidence for betel found in India is the early years of the Christian era, much later than other parts of the region.

Literary sources, however, point to an Indian origin. A Pali text of 504 BC mentions betel (Klebert, 1983). Later, Chinese chronicles of the second century BC describe betel chewing in Vietnam. The next known reference is the Mandasor Silk Weaver's Inscription from India of about AD 473. Areca-nut in Indonesia was mentioned in a Chinese chronicle of the first half of the sixth century (Book 54 of the *History of the Liang Dynasty*). Persian descriptions of betel chewing appeared in Indian literature of the eighth and ninth centuries.

From the tenth century onwards, literary sources provide plenty of evidence that betel was widely used in the region. Champa (Vietnam) gave tribute to China in the form of areca-nuts in the tenth and eleventh centuries (Wong, 1979). The stele of King Ramkamhaeng, of the Sukhothai Kingdom in Thailand, purportedly written at the end of the thirteenth century, says 'The people of this land of Sukhothai ... celebrate the Kathin ceremonies ... with heaps of areca nuts.' An Angkorian inscription mentioned that 'betel nut' was the most important food given to inmates.

The earliest European reference to betel was made by Marco Polo in the thirteenth century, who noted that the people of India always have a quid in their mouths. Other early travellers, such as Ibn Batuta and Vasco Da Gama, also observed betel chewing in the East.

In addition to material evidence, the oral traditions of at least two areas of South-East Asia give insight into the origins of betel. The symbolical use of betel in Cambodia can be traced to a legendary Prince Prah Thong who marries a serpent princess. She gives the prince a betel quid as a pledge of her trust, and since this time betel has been used to bond relationships in Cambodia.

A folk-tale of a mythical king relates the origin of betel chewing in Vietnam. Twin brothers, Tan and Lang, fall in love with the same beautiful girl. Since they are devoted to each other, one of them agrees to let the other marry her. Then, one day, the wife accidentally touches the hand of her brother-in-law which angers her husband. The brother-in-law is so distressed over the incident that he runs away. When he reaches the bank of a stream he dies of grief and the gods turn his body into a white limestone rock, symbolizing his devotion. The husband is upset at the absence of his twin brother and goes to look for him. When he reaches the stream he sees the fate of his brother. He is so grieved that he dies in the same place and turns into an areca palm. Finally, the bride goes to look for the two brothers. When she reaches the river bank she meets the same fate and turns into a betel vine which grows beside the rock and entwines itself around the palm tree. In spite of a drought, the palm and vine remain green. The king hears of this and orders them to be brought to him. He places both in his mouth and is engulfed by a feeling of well-being. Ever since, betel has been chewed in Vietnam.

From these sources—linguistic, archaeological, literary, and oral—it seems likely that betel chewing was practised in South-East Asia in prehistoric times. From the beginning centuries of the Christian era its use spread throughout the region, and from the tenth century onwards, it appears betel has been used regularly.

2
The Quid

THE composition of a quid can be varied by the choice of the ingredients and the way they are combined. The most popular method in South-East Asia is to put some lime paste on the leaf and add thin slices of an areca-nut. Then the leaf is folded, like wrapping a present, to the desired shape and size. (A popular belief is that the character of a person can be judged by the way he or she folds a quid.) Finally, the wad is placed between the teeth and the cheek and pressed with the tongue to allow sucking and chewing. In some parts of the region, though, the ingredients are placed in the mouth one at a time. Yet another way was noted by Albert S. Bickmore in the nineteenth century: 'The roll [quid] is taken between the thumb and forefinger, and rubbed violently against the front gums, while the teeth are closed firmly, and the lips opened widely. It is now chewed for a moment and then held between the teeth and lips, so as to partly protrude from the mouth. A profusion of red brick-colored saliva now pours out of each corner of the mouth....'

The Ingredients

The Nut. The so-called 'nut' used for betel chewing is actually a seed of the *Areca catechu*, a member of the palm family. (This book uses the generally accepted term of 'areca-nut' when referring to the actual seed of the areca palm.) Its distinguishing features are a slender trunk about 20 centimetres in diameter, with a cluster of leaves at the top sheltering stalks of nuts (Colour Plate 2). Amongst the tallest of the palms, the *Areca catechu* reaches a height of 12–16 metres. A grey, fibrous bark surrounds the trunk. These characteristics, combined with graceful proportions, set it apart from other members of the palm family. 'A stem as straight and beautifully shaped as the shaft of a Corinthian pillar,' observed

William Alexander, an English traveller in 1805.

Areca catechu is grown from a seed. It requires little attention to cultivate and needs only a humid atmosphere, a perennial hot climate, and damp soil. It therefore grows better in coastal, rather than inland, areas. Sumatra, western Java, Borneo, the Malay Peninsula, southern Thailand, southern Burma, and the Philippines are the main areas in South-East Asia for the cultivation of the *Areca catechu*. The palm reaches maturity in 1–2 years; after another 5–6 years it starts to bear nuts, producing 200–800 annually for the next twenty years. Plate 6 is a contemporary drawing of a stalk of areca-nuts. White blossoms precede the nuts 'emitting a most fragrant scent at a considerable distance especially mornings and evenings', wrote Nieuhof.

The areca-nut, the only part of the palm used for betel chewing, is round to oval in shape and about 5 centimetres long at maturity. When the nut appears, it is green with a smooth exterior, but it gradually turns yellowish to brownish with a tough, fibrous husk. The areca-nut is 'a kind of great Acorn, which yet wants that wooden Cup wherein our Acorn grows ...', remarked de la Loubère. The interior of the nut consists of a white pulp with a brownish-orange core, which is the part used for betel chewing (Colour Plate 3; Plate 7).

Both ripe and unripe nuts are chewed and connoisseurs have

6. A stalk of nuts from the *Areca catechu* palm. (Contemporary drawing by Yanyong Manoopol)

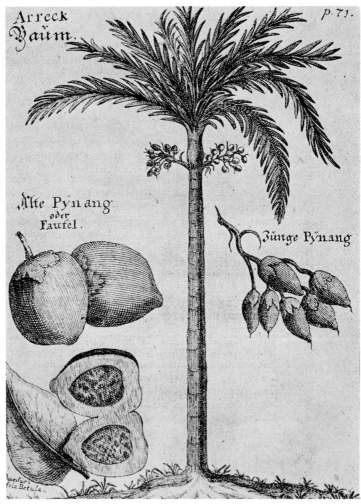

7. Areca–nut and *Areca catechu* palm, from Georg Meister, *Der Orientallis Indische Kunst-und Lust-gartner*, Dresden, 1692. (Courtesy Antiques of the Orient, Singapore)

distinct preferences. Some prefer the tenderness of a nut picked before it ripens, contending it is succulent and sweet-tasting. 'Betel nut is most esteemed when it is young before it grows hard, exceedingly juicy because this encourages spitting,' wrote William Dampier, the first Englishman to reach the Philippines. Others prefer a ripe nut. At maturity 'it is always very bitter and savory', stated de la Loubère.

The nut is used either in its natural state or cured. Generally, people living in humid climates prefer a raw nut, whereas those living in drier areas or parts of the region where the nut is scarce, chew the cured variety. One method of curing is to peel the husk and then boil the seed in water. Another is to dry the nut in the sun until the husk shrinks and then remove the seed. A popular method in Malaysia is to slice the nut in half and then dry it in the sun. A variation of the sun-dried method is to smoke the nut with benzoin. Yet another way to cure the areca–nut, especially if a supply is not readily available, is to store it in salt for two or three months.

The *Areca catechu* palm provided a lucrative source of income for Thailand in the nineteenth century, when all 'fruit-bearing' trees were taxable. The Kamthieng House, a traditional teak wood building, belonged to a family from northern Thailand, who were in charge of administering the collection of tax on all areca-nut trees in the area. When the house was moved to its present location in the grounds of the Siam Society in Bangkok, areca-nut trees from the original site were also transplanted as it was considered unlucky to leave them behind. These trees can be seen standing beside the Kamthieng House in Colour Plate 2.

The Leaf. A leaf from the vine of the *Piper betle* pepper plant is used for betel chewing. The bright green leaf is broad—about 15 centimetres—with defined points (Plate 8). It looks 'like a citron leafe, but longer, sharper at the end', thought van Linschoten. 'The leaves are marked down the middle by pretty ribs or veins, with five or six down the side, and hang from crooked stalks bent low, about as wide as a finger ...' observed John Heydt, a Dutchman, in the middle of the eighteenth century.

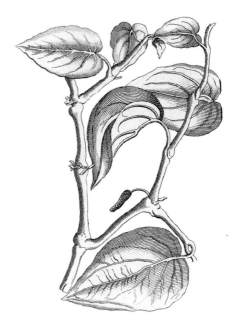

8. Leaf of the *Piper betle*, from Antoine François Prevost, *Histoire Generale des Voyages,* Paris, 1758. (Courtesy Antiques of the Orient, Singapore)

The custom of keeping the right thumbnail long for removing the centre vein of the leaf originated with a legendary Indian monarch who was poisoned by a hair hidden in the central vein of a betel leaf. Even today, in some areas, the main vein of the betel leaf is removed. Snipping the tip of a betel leaf is also customary in parts of South-East Asia because the tip is believed to contain medicine from the gods. A favourite legend tells of a young prince who has an incurable illness. He sees a serpent with a betel leaf in its mouth and, knowing it is a gift from the gods, takes the leaf and removes the tip to avoid swallowing the serpent's venom. He chews it and is cured.

The betel vine is cultivated from cuttings. It requires rich soil and a warm dry climate, and plenty of attention. It also likes shade and, as with other vines, is usually trained to grow up another tree or pole for support and protection from the sun (Colour Plate 4). Betel chewers contend the support chosen for the vine affects the taste of the leaf. A kapok tree or a coconut palm are purported to be ideal supports for producing a tasty leaf. In central Burma, tamarind and plantain trees are favoured to support the betel vine. Occasionally, the betel vine is planted on raised beds, like hops, and trained to grow on wooden slats as seen in a drawing published in Nieuhof's book (Plate 9).

Like the areca-nut, the betel leaf is used in both the unripe (green) and ripe (yellow) stages. There is a distinct preference amongst connoisseurs for the unripe leaf when it is a dark green colour.

9. 'A Betell and Pinangh Garden', from Johan Nieuhof, *Voyages and Travels to the East Indies 1653–1670*, London, 1704. (Courtesy Antiques of the Orient, Singapore)

The Lime. Lime for betel chewing is obtained from various sources, depending on availability. To make it suitable for chewing, the lime is ground to a powder (calcium oxide) and mixed with water to a paste-like consistency (calcium hydroxide). In this form, it is called slaked lime and is white in colour. Limestone chalk (calcium carbonate), obtained from mountain lime, is used in Thailand, Laos, and Vietnam. Sea shells and molluscs, such as snails, provide sources of lime in the island areas. Mussels and other freshwater shellfish from rivers and streams are used in the Philippines. Coral provides the source of lime in parts of Indonesia. Tavernier, a Frenchman, told of an opulent source of lime during an audience with the King of Bantam (Java) in 1648: he saw a woman sitting beside the king 'who held in her hands a small mortar and pestle of gold, in which she dissolved seed pearls'.

Lime is pulverized in different ways, depending on its origins. Sea shells are burned and then crushed with a hammer in the Philippines; in Indonesia shells are crushed with the hands. After reducing them to a fine powder, water, and sometimes a bit of coconut oil, is added to form a paste. A similar method is used in the village of Ban Phluang, in north-eastern Thailand, with the addition of cumin or turmeric which gives a pink or reddish cast to the lime paste.

Lime for betel chewing is almost always kept in a separate container. The reason why lime containers are so common is unknown. It may have derived from an ancient Malay belief that poison could be added to the lime paste.

'If you eat lime you will have a stomach ache', warns a Thai proverb. This is based on the belief that long ago a small lizard climbed down the wall of a house and stole some lime from a woman's betel basket. Because of this crime, the lizard still croaks, proclaiming his guilt. Children are told that this lizard will come down the wall and eat their livers if they cry too much.

Other Ingredients. Additional ingredients in a betel quid are a status symbol. The greater the number and the more exotic, the higher the host's prestige. In addition to the three basic ingredients for a betel quid, a Sanskrit text lists cardamom, clove, camphor, musk, nutmeg, copra, black pepper, and dry ginger. Not surprisingly, most of these are spices which abound in the region and were among the riches of the Moluccan islands that lured Europeans to the East in the sixteenth century. The local peoples merely added spices to the betel quid just as they did to other foods. Cloves and anise seed give a pleasant aroma to the chew. A stick of clove is used to secure a folded or rolled quid. Cinnamon, coriander, ambergris, and nutmeg add flavour and thus enhance the taste. Cardamom stimulates the flow of saliva.

'Borneo camphor' was mentioned by a Chinese traveller, I'Tsing, in the seventh century as one of the ingredients chewed with areca-nuts at a feast in Sumatra. However, it was only used by the 'richer and mightier' (Burkill, 1935).

Tobacco is a modern addition to the quid. It is combined in various ways but always to give flavour to the quid. Sometimes shreds of tree bark are substituted for tobacco. In Malaysia, shredded tobacco is rolled into a wad and placed in the jaw during chewing. In north-eastern Thailand, a piece of tobacco is moistened and placed in the jaw. A pinch of finely shredded young tobacco is often used to wipe the lips before and during chewing. Shredded tobacco leaves are sometimes rubbed on the teeth before chewing the quid; afterwards, the teeth may be cleaned with a tobacco leaf.

Substitutes for the main ingredients of a betel quid are used in some areas. For instance, gambier, an astringent extract obtained from the leaves and young shoots of the *Uncaria gambir* plant and also a tanning agent, has been a popular alternative in Malaysia since the fourteenth century and is also used in Indonesia. Because it is cured before use, gambier is not perishable. Boiling water is poured over the leaves and the juices squeezed out. As the mixture cools, it crystallizes. Then it is moulded into small balls or squares. Gambier 'is somewhat insipid, having a peculiar gummy kind of taste', wrote Ambrose B. Rathborne, an English visitor, when he tasted it in Malaysia in 1898.

Trade of the Betel Chewing Ingredients

Betel chewing has supported an active regional trade since the early years of the Christian era. Chronicles record that areca-nuts were sent from Java, Champa (now Vietnam), and Burma to China sporadically from the first century up to the thirteenth. However, records on the trade of betel chewing ingredients are irregular until the last century, although Europeans did comment on the extent of the trade. Cort, for example, wrote that the 'yield [of betel] in Siam is immense, and great cargoes are shipped to India and China, where the same disgusting habit prevails'. In the early nineteenth century, gambier was the 'most important article of Singapore produce', according to the *Straits Settlement Records* of November 1836. Plantations, which were always operated by Chinese immigrants, were a main agricultural industry.

The island of Penang was a main centre for the trade of betel chewing ingredients and even owes its name to the custom. The English word 'penang' derives from the Malay name for the areca-nut palm, *pinang*. In the nineteenth century, Europeans were acting in Penang as middlemen in the trade of ingredients for betel chewing. They purchased areca from native growers and sold it for as much as 1–200 per cent profit.

Even today, the demand for betel chewing ingredients creates an active internal trade. The methods of supplying and marketing these products are much the same everywhere and have changed little over time. A typical vendor in a rural market—characteristically a woman—is seen in Colour Plate 5. Here, the vendor is sitting on a woven mat with her wares spread out in front of her—a stack of green areca-nut stalks on one side and a pile of betel leaves on the other. Her lips are, of course, stained red from betel chewing. It is said in rural Thailand that most women go to the market daily just to buy fresh betel leaves. 'A Man Selling Betel' is depicted in an early nineteenth-century aquatint (Plate 10). Wearing traditional Chinese dress, he displays his goods on a wooden stand and sits on a stool made of bamboo, cutting the areca-nut with a knife. In Malaysia, betel ingredients

10. 'A Man Selling Betel', from William Alexander, *Costume of China*, London, 1805. (Courtesy Antiques of the Orient, Singapore)

are marketed by special traders, usually women, who function as secondary dealers selling to rural markets. Female betel vendors, using small boats as their stalls, are a common sight along the banks of the river at Bangkok. These women are 'amongst the most polite and obliging saleswomen in the world', observed Frederick A. Neale, a nineteenth-century visitor.

Betel Chewing and Health

Chewing betel evokes a mild euphoria, and it is this general feeling of 'well-being' that contributes to the popularity of the custom. The ingredients of the betel quid, though, are not narcotic and betel chewing is not addictive although it can be habit-forming.

The properties of the areca-nut relevant to betel chewing are alkaloids and tannin. The main alkaloid, arecoline, is toxic and has a stimulating parasympathetic nervous action, giving the betel chewer a relaxed feeling. This alkaloid activates secretion, increases smooth muscle activity, salivation, and thirst, but reduces appetite. It gives a red colour to the saliva, teeth, and faeces. The alkaloids in the areca–nut also contribute nitrogenous matter to the diet which neutralizes stomach acids and acts as an astringent. The tannin in the nut contributes the property of astringency.[1]

Areca–nut is widely used in veterinary medicine, mainly to expel parasitic worms in animals. The pulp of the nut is used for relieving pain in the stomach of humans. As an astringent it hardens the mucous membranes of the stomach. In Malaysia, young shoots of the *Areca catechu* palm are believed to be effective in aborting a pregnancy. The root of the palm is given to cure dysentery. In Malaysia, too, the *Areca catechu* flowers are put into the bath water of a woman who has just given birth.

The *Piper betle* leaf contains phenols which contribute to its aromatic scent and pungent taste. It also contains eugenol, a clove-oil compound, which is a powerful natural antiseptic. This role as an antibacterial agent accounts for its effectiveness in curing infections, especially of the skin and the eyes. Local peoples also use the juice of the leaf to aid in the healing of headaches and fever, while stalks of the betel vine are used for glandular swellings.

According to the universal classification of food, the areca–nut and the betel leaf complement each other and are, therefore, in harmony. Since the areca–nut is 'hot' and the betel leaf 'cool', they act together to keep the human body in balance. Some claim that the areca–nut is an aphrodisiac, perhaps because of its classification as a 'hot' food. Conversely, the betel leaf, as a 'cool' food, is believed to relieve 'hot' illnesses such as headaches and fever.

The relationship between betel chewing and oral cancer is unclear. In some areas where betel chewing is concentrated, a

[1]A recent report in the medical journal *Lancet* (10 May 1992) reported that a substance in the areca–nut narrows the bronchial tubes and can 'provoke severe asthma attacks'.

high percentage of mouth cancer is reported. These claims, however, are not supported by research. Scientists from the Cancer Research Institute in Gujarat, India, have however warned that chewing betel can cause genetic damage which may affect the children of users (*Bangkok Post*, 19 April 1989).

The effect of betel chewing on the teeth has not been determined. It does, though, turn the teeth red, and if betel is chewed over a prolonged period without cleaning the teeth, they will turn a black colour. Chan, a fisherman who lives on the edge of the Tonle Sap Lake in Cambodia, was talking with a European visitor in the late 1920s when he noticed the man's lips 'parted in a ready smile'. He was startled to see the man had teeth—'clean white teeth like those of a young child'—and with no trace of 'betel-juice' in his greying beard. 'Truly a marvellous person,' thought Chan. This encounter illustrates the uniqueness of white teeth in South-East Asia. Likewise, Europeans were struck by the appearance of black teeth among some peoples of South-East Asia. 'It seemeth that the lips and teeth are painted with blacke blood ... ,' wrote van Linschoten. Major McNair also noticed that 'The effect [of betel chewing] is to stain teeth a dark red, in some cases almost black, and seen in a young girl this is to a European anything but pleasant....'

As recently as fifty years ago, black teeth were considered a mark of beauty by Asians and were therefore a desirable feature. It was believed that only animals had white teeth and, since human beings were superior, it was considered shameful for them to have white teeth like animals. 'Your teeth are white!' scolds an Indonesian mother, a traditional reminder to children that they are no better than animals. 'Black like the wings of the beetle ...' was the cosmetic ideal. The Thai women 'dye their teeth ... a jet black colour.... The darker the teeth the more beautiful is a Siamese belle considered ... ,' observed Neale.

A Malaysian folk-tale praises glossy black teeth:

> Whose the cock that struts so bravely,
> His lips a shore beset with bridges,
> Bridges of black shining palm-spikes,

Teeth as stems so sharp and knitted,
Mouth a boatful of red nutmegs,
Ebon teeth like bracelet circle?

An amorous Siamese sailor sings the charms of a black-toothed damsel in 'The Boatman's Song':

... as for thy nose, I'm certain that
None other has one so wide and flat:
And the ebony's bark, in its core beneath,
Was never so black as thy shiny teeth.

The 'betel chewing look' even made it to Broadway in the musical 'South Pacific' through the character of Bloody Mary:

Bloody Mary's chewing betel nuts,
She is always chewing betel nuts,
Bloody Mary's chewing betel nuts,
And she don't use Pepsodent!
Now ain't that too damn bad!

The popularity of black teeth from betel chewing was so great that dentists in the nineteenth century had to make sets of black false teeth, according to American missionaries (Vimol, 1982).

Prolonged chewing is generally believed to keep the gums healthy by strengthening them. It also seems to prevent tooth decay as long as the teeth are cleaned. The reasons for these positive aspects of betel chewing on teeth are probably the fluoride content and the antibacterial effect of the betel leaf. Surveys in New Guinea and East Java have shown that cavities are markedly less frequent among betel chewers. Gum disease, though, is common because of the irritating effect of the lime. Pieces can become wedged between the teeth causing gaps where food can lodge and attract tooth-destroying bacteria. The teeth may become loose and with prolonged chewing can even fall out. Lime grinds the enamel black and, when chewed, also blackens the dentine. Teeth can also be blackened by a deliberate method of using vegetable dyes.

Interestingly, the opposite opinion of the colour of teeth is held in India, a nation of betel chewers. Pearly white teeth are acclaimed and everyone is required to ceremoniously wash their mouth four times daily.

> Like the white buds of tuberose in a dark night;
> through the lines of betel shone out her white teeth

waxes the lines of a poem in the *Book of Indian Beauty* (Anand and Hutheesing, 1981).

3
The Symbolism

BOTH the areca-nut and the betel leaf have a widespread symbol-ical significance in the culture of South-East Asia. Betel is believed to be instrumental in establishing communications, and the symbolism focuses on two aspects of its power in this area: contacts with the spiritual, or supernatural, forces, and social and sexual relationships between a male and a female. The earliest symbolical use of betel was most likely as a sacrificial offering for animistic worship. Through the centuries betel has been assim-ilated into many parts of the culture and serves as a traditional offering at animistic rituals, festivals in the lunar calendar, and Buddhist and Hindu ceremonies. The history of the symbolism of betel has been compiled from information passed down orally through the generations and all of the customs described in this book form a part of the history although, in some cases, they are no longer practised.

Symbolism and the Spirits

The ancient belief that every object has a spirit created the need for propitiation. All spirits, regardless of whether they are good or evil, must be dealt with and controlled through rituals. Offerings of betel are made to satisfy, win over, or thank good spirits and to exorcize evil ones. Colour Plate 6 shows a villager in north-eastern Thailand making a typical offering of betel to a miniature house which has been provided for the spirit *Phra Phum*, Lord of the Land. It is believed that if he is taken care of through appropriate offerings he will guard and protect the people who live on the land near the spirit house.

Spirits of the land and water are given special attention in agri-cultural areas where adequate rainfall and fertile soil are essential for the cultivation of rice. The Water Festival, which marks the beginning of the Buddhist year in Thailand, is a joyous celebration that coincides with the full moon of the lunar calendar. Offerings of betel are made to the water spirits asking for plentiful rainfall in the forthcoming year. After the rainy season in Thailand, on the night of the full moon of the twelfth lunar month, thanks are offered to the spirits of the water by floating graceful boats made from banana leaves on the waterways. Betel, flowers, incense, and candles traditionally fill the boats. To invoke rain in Malaysia, the spirits of the water are contacted by planting betel in a dry area followed by a rain-calling procession. In Thailand, the spirits of the soil are propitiated with offerings of betel at a Ploughing Ceremony. In Malaysia, fishermen hang offerings of betel, meat, and vegetables at the prow of the boat to the spirits of the land and sea.

Evil spirits are the most feared of the supernatural forces because they cause illness, so many rituals focus on exorcizing the evil spirits and replacing them with protective ones. A medium is considered to possess supernatural power in establishing communications between the spiritual and earthly worlds and is especially adept in dealing with evil spirits. There are both male and female mediums, but, in Thailand, females are thought to be particularly gifted. In a typical ritual, a medium sits on the floor chewing betel and chanting with a mound of betel leaves near by. Not surprisingly, betel figures prominently in chants such as this one from Malaysia:

> He sat down cross-legged and reached out for a small betel box.
> Chewed betel twice or thrice then ceased....

(Gimlette, 1971)

Areca-nut is tossed on the bed of a child by the Dayaks in Sarawak to placate the evil spirit, Jirong, and the following verse is chanted:

> Here, take it, Jirong—
> this old betelnut is for you,
> the young betelnut is for us.
> Let the child live a long life and be healthy, live
> until he walks with bent back, leaning on a low tongs-staff,

leaning lower on half a coconut shell,
and then grasping the bamboo pipe for blowing fire,
creeping close to the ground.

(Rubenstein, 1985)

The Dayaks used to recite another chant to propitiate the spirits accompanying warriors into a head-hunting battle. As the warriors rest, they chant:

Sitting on the wood and quickly reaching
for the colored betelnut basket.
Pulling out a long pointed knife made with lead
for splitting a young betelnut. To tear the betel leaves,
yellow as the durian leaf and the yellow tail of the *ichah* bird.
The vine of the betel leaves seems to crawl
on top of the *majau* tree growing on the mountain.
To pour the lime powder,
which looks like magpie dung beside the track.
The lime powder was made by a Malay man;
the lime powder was made by a wife.
She roasted it one day during the hot dry season,
while in the middle of overgrown secondary jungle.
Chewing the quid makes the lips hot,
red as the ripe *kirumi* fruit near the secondary jungle.
It is spat out and dries like the dried leaves
that float during the days of the hot dry season.
After chewing betel leaves, let us smoke tobacco.

(Rubenstein, 1985)

Betel spittle is considered especially powerful in dealing with illness. If transmitted by a medium, its ability to exorcize supernatural forces is unlimited. Gazing into a bowl of betel spittle, a medium chews a quid and receives an omen. Then she sprinkles the juice of her quid over the body of a person who is ill. She ejects the quid on the back of another patient. Both are cured. Spittle is also used for protection from illness. A midwife, for example, spits the saliva of her quid on to the stomach of a newborn baby.

Felix Chia, a Straits-born Chinese, in his book, *The Babas*, recalled when he was five years old being 'rained' on with drops of red-stained betel saliva by a family employee who had certain spiritual powers: 'With her eyes closed and the *siray* [betel] frothing in her mouth ... she began to blow hard at my face. Then little drops of red liquid, bits of leaves and God knows what else splattered all over my face and the front of my shirt!'

A pregnant woman is vulnerable to evil spirits from the time of conception until delivery and so she must be protected from them. In Malaysia, a ritual is conducted to determine if the delivery will be an easy one for the expectant mother. In one version of the ceremony, a midwife turns a betel tray upside-down. If the contents fall out together, an easy delivery will follow. Another ritual using betel takes place in the seventh month of pregnancy to determine the sex of the baby. A midwife throws small pieces of areca-nut on the floor. The greatest number of pieces lying uppermost, either flat or round, decides whether the baby will be a girl or a boy.

A custom amongst the Dayak tribe is to sprinkle the new-born child with areca-nut and other auspicious symbols, then wrap it in a cloth and lay the baby on a bed of areca-nut palms. After birth, this palm and a betel vine are planted side by side for the child.

In Cambodia, after giving birth the mother must offer betel to the midwife or she will have to follow her for many future lives and the midwife will never turn around to help the mother. Here, and in north-eastern Thailand, the new mother undergoes a 'lying by the fire' ritual. Since fire is considered a purifier, the mother lies on a bed heated below by a charcoal fire for seven days after the birth to dry out the womb. During this time, the protective spirits are assuaged with traditional offerings of betel, flowers, food, candles, and incense.

The symbolical association between betel and spirits makes its use widespread in rituals of death. It is customary in parts of South-East Asia to provide the deceased with appurtenances from the worldly life to accompany them to eternity. The important position of betel on earth makes it an essential item to go with the deceased on the journey to the spiritual world. This belief was practised as early as the sixteenth century on the island of Luzon in the Philippines when betel juice was used to embalm the dead.

The use of betel for funeral rites is also believed to pave the way for a better incarnation for the deceased. At a royal cremation in Bali, betel leaves are amongst the gifts presented to the regent. A funeral in Thailand ends with a social gathering of the mourners who talk, chew betel, and play games throughout the night. The prevailing atmosphere of gaiety at a funeral reflects the Buddhist belief in rebirth and the continuity of life. When a person dies in Cambodia, ritual objects are arranged around the body. Immediately after death, a candle is lit, which is later used to light the funeral pyre. A betel leaf is placed between the fingers of the deceased and a fig leaf inscribed with a verse is put on the lips. Formerly in Burma, it was a custom to offer a dying man a betel quid and a cup of water. Today, when a man asks for these two things, it is understood he feels there is nothing left to live for. A man condemned to die in Malaysia is given a betel quid to assuage his soul. A Malay proverb teaches that an early death can mean the promise of a fruitful life: 'The prop was snapped asunder as the betel-vine ascended'.

It is the duty of those living on earth to honour and propitiate the spirits of their deceased ancestors. Betel quids and rice are typical offerings used for the rites associated with ancestors.

The symbolical use of betel that began with offerings to spirits was later assimilated into religious ceremonies. Betel, for example, is linked symbolically to the Hindu trinity: the areca-nut to Brahma, the Creator; betel leaves to Vishnu, the Preserver; and lime to Shiva, the Destroyer. Its association with Buddhism is represented in a group of stones in the Maldives known as the Great Mound of Fua Mulaku which holds areca-nut and leaves mixed with lime so that the Buddha can chew betel.

Symbolism and Male–Female Relationships

Betel is considered a significant element in fostering both social and sexual relationships between a male and a female and nowhere is this more prominent than in the language, folklore, and poetry of the region. It has even penetrated the vocabulary as numerous words derived from the equivalent of 'betel' relate to a union between the male and the female. In Malay, for example, compounds of *pinang* (areca-nut) mean 'to court' or 'to propose'. *Meminang* is 'to ask in marriage' and *pinangan* is 'betrothal'. *Pinang muda* is a euphemism for a go-between of lovers and draws a correlation with the ideal areca-nut which has two perfectly matching halves. *Sireh*, the Malay word for betel leaf, means 'a young girl who is eligible for marriage'. *Leko passiko* ('a bundle of betel leaves') is an offer of marriage in Makassar. *Khan mak* ('a basin of betel nut') refers to a wedding in both Thai and Lao. In Thailand today, the phrase means a present for an engagement.

The idea that chewing betel stimulates passion and brings out charm is reflected symbolically in many tales and beliefs involving relationships between a male and a female. Betel is present from the earliest encounter between the two. This connection is mentioned often in Vietnamese literature. A proverb teaches: 'A quid of betel is the prelude to all conversation'. And 'Will you accept a quid of betel and tell me in which village you live?' a male asks a female. Another story recalls a young girl picking mulberry leaves. Two men who were fishing near by spoke to her but she replied, 'My parents have warned me that young girls should not accept betel quids from strange men.' A chant of the Dayaks tells that betel leaf helps a couple 'to tangle and mingle together like coursing water'.

Batak girls, ten or eleven years old, in Sumatra sleep in a communal house with a chaperon. A meeting between a girl and a potential husband takes place in the house and he initiates a conversation by offering betel to the girl. A childhood story related by a Burmese woman, Khin Myo Chit, recalls a similar custom. It was traditional for a young girl of marriageable age to sit at her loom with a betel box filled with fresh ingredients beside her. Groups of potential husbands, known as 'bachelor rounds', called on the girl and engaged in conversation. When she favoured a young man, she prepared a quid and offered it to him, which was a signal for the others to depart. A customary way of courting a prospective bride amongst the Iban people in Malaysia takes place when members of the house are asleep. With betel leaves and areca-nut, he enters the *bilik* or apartment of his prospective bride's family, goes to her bedside, awakens her, and engages in a romantic conversation. If the woman wants to marry the man, she accepts the betel.

A betel quid can be used as an indicator of decisions or the settlement of disputes between males and females. In Java, for example, a woman identifies her preference for a man by the way she folds a betel quid. If she loves him, she sends him a quid wrapped in two leaves with the top sides pressed together; if the undersides of the leaves are pressed together, it means she is not interested. Betel has also been used to confirm the separation of two people as well as to bring them together. For example, formerly in Burma, a wife could ask her husband for a divorce by dividing a betel leaf in half and offering the other half to him; if he accepted and chewed the betel, it signified that he agreed to a divorce.

Ancient legends reflect the symbolism between betel and love. In a twelfth-century Indian verse a man whispers to his lover: 'Dost thou recollect, after passing bits of betel from my mouth into thine, I justly demand them back?' A Thai folk-tale tells about the beautiful daughter of King Traiyatrung who dreamt of a struggle with a snake, a symbolical dream meaning she was going to find a husband. The next day, a man appeared at the palace. The princess sent an offering of betel to him along with a message that if he wanted more betel he should reply by sending her fruits and vegetables, which he did. She continued to send him betel and letters hidden amongst the leaves. Eventually, the prince and princess married and lived happily ever after.

A dramatic scene from the 'Tale of the Pandavas' (the Malay version of the great Indian epic, the *Mahabharata*), relates the parting of Maharaja Salya and his beautiful wife, Dewi Satiysurati. He knew he would never see her again and, before he left, he embraced and kissed his beloved wife, then chewed a quid of betel and put it in her box.

With such a close link between betel and love, it is not surprising that the symbolism extends to erotica. A male woos a female

. Akha hill tribe woman with 'red lips'. (Photograph Martin Madsen)

2. *Areca catechu* palms in the garden of the Kamthieng House, Siam Society, Bangkok.

3. Fresh areca-nut and betel leaves. (Photograph Vance R. Childress)

4. Vine of the *Piper betle*. (Photograph Vance R. Childress)

A Malaysian vegetable seller. (Photograph by Karen R. Childers.)

6. Villager in north-eastern Thailand making a betel offering to the spirits. (Photograph Vance R. Childress)

7. Offerings of betel, rice, and fruit at a wedding in north-eastern Thailand. (Photograph Vance R. Childress)

8. Spreading lime paste on a betel leaf with a spatula. (Photograph Vance R. Childress)

9. Pulverizing ingredients for a betel quid with mortar and pestle. (Photograph Vance R. Childress)

0. Betel set, woven fibre, Thailand. Diameter 23 centimetres. (Courtesy Elephant House, Bangkok)

1. Lime container, bamboo, Timor. Height 15.5 centimetres.

12. Betel box, wood, with painted decoration, northern Thailand. Height 30 centimetres. (Courtesy Industrial Finance Corporation of Thailand)

13. Betel box, wood, with fresh ingredients, north-eastern Thailand. Height 9 centimetres. (Photograph Vance R. Childress)

14. Betel set, wood with brass rim and fittings. Diameter 80 centimetres. (Collection of Mohd. Yunus Noor)

15. Betel box, lacquer, Burma. Height 14 centimetres.

16. Betel box, lacquer, northern Thailand. Height 9 centimetres.

17. Sign of the zodiac decoration on a betel box, lacquer, Burma. (Courtesy Elephant House, Bangkok)

18. Ceremonial betel box on a stand, gilded, Burma. (Courtesy Sylvia Fraser-Lu)

19. Betel box, lacquer with
    silver inlay, Burma.
    Height 23 centimetres.
    (Courtesy Elephant
    House, Bangkok)

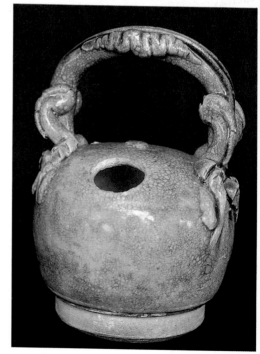

20. Ceramic lime container,
    Vietnam. Height 17 cen-
    timetres.

21. Betel cutter, steel with silver inlay, Burma. Length 13 centimetres. (From the Samuel Eilenberg Collection; courtesy of the owner and his publisher, Hansjörg Mayer)

22. Ceremonial betel box on a stand, silver, Burma. Height 25 centimetres. (Courtesy Elephant House, Bangkok)

23. Lime container, silver, copper, and horn, Laos. Height 18 centimetres. (Collection of Donald L. Petrie and Niyanee Srikanthimarak)

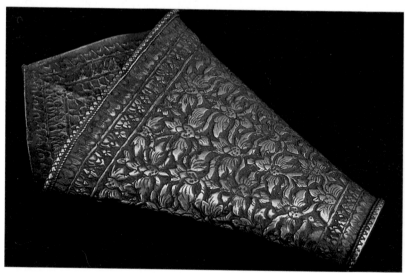

24. Betel leaf holder, silver, Thailand. Length 12 centimetres. (Collection of Donald L. Petrie and Niyanee Srikanthimarak)

25. Betel set, niello, Thailand. (Collection of the late Prince and Princess Chumbhot of Nagara Svarga)

26. Betel set, gold, Thailand. (Courtesy River City Auction House)

in an evocative Vietnamese poem (quoted in Milner, 1978) with betel symbolism:

I enter my garden to cut a fresh areca nut
I divide it into six and invite you to chew a quid of betel
This betel is prepared with Chinese lime
And I have added a little campanula in the
middle and some spicy cinnamon at each end
This betel will truly intoxicate you
Whether it be strong or mild, spicy or hot
Whether or not we become husband and wife
I invite you to eat two or three quids to calm my excitement a little.

Betel was listed as a necessary adjunct to sex in the Kama Sutra. An early Indian text instructs that a young and beautiful woman cannot go to meet her lover in a secret place without 'golden necklaces and betel leaves and flowers'.

In India, men eye women amorously and they 'make their beginnings of love ... for that both night and day they do practise nothing else but make it their [only] worke, and to make nature more lively [to abound and] move them thereunto, they do use to late those Bettles, arequas, and chalk ...', wrote van Linschoten.

Betel has an explicit sexual symbolism in some parts of the region. In Vietnam, for example, the vine of the betel leaf (vagina) wraps around the areca-nut (male) with lime at the base which, when ground, produces a lime paste (male and female) which dresses the leaf and the nut, and a poem describes 'breasts pointed like areca nuts'. A Malay proverb draws an analogy between a well-matched couple and the cleft of an areca-nut.

Betel has been closely associated with ceremonies involving marital union since ancient times. It was offered as a prelude to discussions of partners, dowries, and other arrangements necessary for a marriage. Acceptance of the betel signified agreement to the proposal being discussed. For example, the parents of a pro-spective bride acknowledged the engagement of their daughter by accepting a betel quid offered by the husband-to-be. During betrothal negotiations amongst the Malays, the parents of the groom offered a betel tray to the parents of the bride. If it was turned upside-down, it meant the proposal was not accepted; but if the betel tray remained upright, it signified that the arrangement was agreeable to both parties.

During the engagement ceremony of a Malay couple, neatly arranged betel leaves, the ring, and other gifts are carried to the home of the bride. The ring is placed in a betel tray resting on a fan of betel leaves.

A betel box is often part of the bride's dowry. In north-eastern Thailand a carved wooden betel box is the traditional gift from a bride to the groom. In Kelantan, the dowry is delivered on the wedding day and the groom does not join in the procession to the bride's house until he receives the betel box. In Perak, the initial engagement offering consists of 20–40 betel quids and two sliced betel nuts.

Offerings of betel and food to the monks by relatives and friends is a gesture of making merit preceding the wedding ceremony in Cambodia. The bride and groom throw betel leaves at each other during the wedding ceremony in Java. A 'betel tree' is carried in the wedding procession in Sumatra. The groom's mother chews betel in the house with the bridal couple on their first night together in Timor. A marriage ceremony begins with a procession of relatives and friends carrying gifts, including betel trays, in Cambodia. Monks sit around an altar which is adorned with flowers, fruit, betel, and incense.

According to a wedding custom of the Straits-born Chinese in Malaysia, the groom 'pays' the parents of the bride for raising his bride a virtuous girl. The ceremony for this gesture is accompanied by a container of betel leaves with gambier, slaked lime, and areca-nut. Amongst the Lingga Dayaks and the Balu of Borneo, the traditional wedding gift, the areca-nut, is offered in a ceremony called the *bla pinang* which means 'division of areca-nuts'.

A 'washing of feet ceremony' is held in Thailand on the day of the wedding. A matron of honour feeds the bride and groom rice, fruit, and betel to ensure happiness, peace, and harmony in their marriage. These offerings are seen at a traditional wedding ceremony in north-eastern Thailand in Colour Plate 7. The following day a *bunga gomba* ritual takes place. A piece of white cloth is held over the heads of the bride and groom. Ritual water and slices of areca-nut are poured on the cloth to keep away bad luck and obstacles.

The marriage ceremony of the Dayaks in Sarawak is celebrated by splitting an areca-nut into seven pieces which are put on a brass tray with seven betel leaves, seven pieces of gambier, and small lumps of lime. Relatives and friends gather around the tray and share its contents and discuss the binding nature of the marriage contract. A Malay custom is to place a betel box beside the marriage registrar and a betel tray on the dais during the 'sitting-in-state' ceremony. The bride is given an areca-nut that has been blessed with auspicious charms to increase her radiance.

The wedding ceremony in Cambodia includes a lime container and a knife to cut the areca-nut. Upon retiring after the ceremony, the bride enters the chamber first with the end of her scarf held by her husband. Inside they exchange and chew betel quids together.

It is customary to place a betel box outside the bridal chamber on the night the bride and groom consummate their marriage in Thailand. If the box is overturned by the groom during the night, the bride's virtue is in question and a family inquiry follows. A similar post-marriage ceremony performed in Malaysia and Thailand is for the preparation of the bridal bed. It is conducted by a respected elderly married couple who place rain-water symbolizing purity, and candles, incense, and betel on the altar.

4

The Art

IMPLEMENTS for preparing, serving, transporting, and storing betel ingredients represent an art as distinctive as the custom itself. Betel sets are made with caring and talented hands from the materials available. The religion, art, and nature of each area provide inspiration for the decorative motifs which add colour and individuality to the forms.

A receptacle, either a box, tray, or basket, to hold the ingredients for chewing betel is essential to a basic set. This receptacle may be divided into compartments or hold individual covered containers made of a different material. The boxes for ingredients vary in shape and size. Besides boxes in geometric shapes, others are crescent-shaped and can be attached to the waist for transporting betel. Some are triangular-shaped holders for rolled betel leaves. A cutter for slicing the nut completes the basic betel set. Generally, in areas where the soft unripe nut is eaten, a knife is used for cutting, but scissor-like cutters are used in areas where the dry or cured nut is preferred. Other accessories include a spatula for removing the lime paste from its container and spreading it on the leaf or putting it in the cheek (Colour Plate 8), a spittoon, and a mortar and pestle for pulverizing the nut to make it palatable to toothless people (Colour Plate 9). The betel paraphernalia is depicted in a nineteenth-century drawing published in Batavia, the headquarters for the Dutch East India Company (Plate 11 and Cover Plate).

Materials used for making betel equipment are closely connected to nature, and countries with similar geographical and climatical features tend to use the same materials and methods. The lacquerware craft, for example, spreads from Burma to northern Thailand where sap-producing and other types of trees necessary for lacquer grow. The material a set is made of depends on the wealth and status of the owner and on whether it is for individual or communal use. The main types of materials used for betel utensils in South-East Asia are natural forms, fibre, wood, lacquer, clay, metal, and shell. Sometimes several materials are used on a single set. The Kenyah in Borneo, for example, weave colourful beads into striking geometrical patterns for betel sets, such as the wooden rectangular box with round covered brass boxes inside for the individual ingredients shown in Plate 12.

Natural Forms. Betel utensils made of natural forms are common in the island areas of the region. The materials are durable, impervious, and lightweight. For instance, lime containers are made from gourds on the Indonesian island of Timor, from coconut husks further west, on Flores, and from carved horn in south-east Sulawesi (formerly the Celebes). Buffalo horn or coconut are also used for spatulas. The Ifugao peoples of the Philippines make lime boxes from human bones, decorating them with pictorial scenes showing the reason for the death of the 'former owner'. Deer horn is used in Burma for cracking the areca-nut; the nut breaks into small pieces when pushed through from the wider to the narrower end of the horn.

Fibre. Reeds and fibres from the tropical rain forests are woven into durable receptacles for betel chewing utensils and skilfully decorated with attractive designs. Because these baskets are lightweight, they are most suitable for carrying betel from place to place, and are particularly favoured by women for transporting betel between the home and the rice-fields. Palm strips are used in west Borneo to weave a betel receptacle suspended from a carrying strap. Reeds of the *Pandanus* plant serve the same purpose in Indonesia.

A fine example of a woven betel set is a round covered box from Thailand. It is intricately made from strips of twisted reed. The interior of the box is divided into four compartments, each bearing a smaller, round, covered box (Colour Plate 10).

Bamboo is an ideal material for small betel receptacles because of its shape, strength, abundance, and hard exterior. It is traditionally used for lime containers in the island areas of South-East Asia. The natural trunk of the bamboo is a ready-made container. The long, narrow cylinder only needs a small plug for the base and a

GAMBAR-GAMBAR
akan Peladjaran dan Kasoekaan Anak-anak dan Iboe-bapanja.

NEDERLANDSCH-INDISCHE PRENTEN.

Tampat toedah.
Kwispedoor.

Tampat daoen.
Bakje voor de bladeren.

Tampat kapoer.
Kalkbakje.

Tampat pinang.
Bakje voor pinang.

Tampat sirih orang ketjil.
Beteldoos voor minderen luiden.

Kebon sirih.
Bladtuin.

Tampat gambir.
Gambirbakje.

Pinang.
Pinangnoot.

Daoen.
Pinangblad.

Katjip.
Pinangschaar.

Toetoepeja.
De deksel.

Tampat sirih.
Beteldoos.

Tampat sirih.
Beteldoos met tootsboeren.

Makan sirih.
Betel kauwen.

Digambarken olih
TOEWAN KOLFF,
jang dahoeloe di negeri Betawi.

Hoogt ± 0,3 Meter.

11. 'Gambar-Gambar akan Peladjaran dan Kasoekaan Anak-anak dan Iboe-bapanja' by Toewan

42

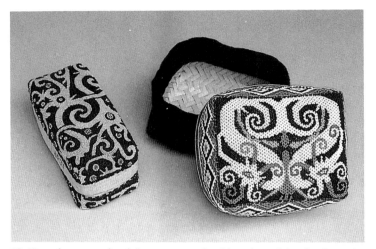

12. Kenyah woven betel boxes covered with coloured beads, Sarawak. (Courtesy of the Sarawak Museum)

slice through to the inner core for a lid. After sanding the surface to a smooth finish, the cylinder is ready for decorating. An intricate design of stylized geometrical forms is cut into the bamboo with a sharp tool, then a dark-coloured vegetable dye is applied to the incised area for contrast and definition. The final result is a pleasing interplay of light and dark (Colour Plate 11). The same look is achieved on the island of Luzon and in Papua New Guinea by burning the incised area with a hot knife which turns the decoration black. The islanders of Mindanao prefer a more colourful design made by cutting a pattern into a piece of coloured paper and then pasting it on to the bamboo. The tube is wrapped in banana leaves and boiled. After rinsing, the cylinder is polished with the pulp of a young areca-nut. The Batak tribe of Sumatra use a similar type of container but add a stopper made of wood or horn.

Rattan from the stems of a *Calamus* palm is used for a special type of lime container in the Philippines. The unique feature of this rattan cylinder is a sifter for shaking powdered lime inserted in one end. Another type of lime container made by the male

43

members of the Hanunoo peoples of the Philippines is a wooden tube carved in the shape of a phallus; perhaps the craftsman hopes the symbolic sexual prowess of the phallus will be transferred to him.

Wood. A betel box made from the wood of the teak forests in northern Thailand is unique to that area. The simple yet functional receptacle is square or rectangular and usually supported by a pedestal. One half of the interior is divided into small compartments for the ingredients; the other half is for the betel leaves, cutter, lime container, and mortar. On earlier examples, the box is secured with bamboo pegs by using the tongue-and-groove method of construction; nails are used on more recent ones. A fret pattern and carved designs are typical decoration. The box is finished with a wooden trim and brightly coloured paints, lacquer, or a dark stain (Colour Plates 12 and 13).

Plate 13 shows a typical betel box most commonly used on the Malay Peninsula for rites of passage. It is a rectangular wooden

13. A typical Malay wooden betel box covered with gold-embroidered red velvet.

box with internal compartments and covered—on the sides and top—with gold-embroidered red velvet.

An exceptional use of wood is a betel set from the archipelago. The round container, hand-carved from the bole of a tree, reveals the fine, dark-veined grain of the wood. A band of bronze around the rim is boldly impressed with Arabic script, the religious language of the Muslims. The boxes, leaf holder, and cutter are made of bronze (Colour Plate 14).

Lacquer. The craft of lacquerware has a long tradition in Burma and northern Thailand and each country has, at one time or another, influenced the other. Lacquering is a practical way of strengthening an object and making it impervious to liquids. This is particularly useful for keeping betel ingredients dry during the rainy season. This feature and the availability of appropriate woods make lacquer a popular and widely used material for betel containers of all sizes and shapes.

The process of making a lacquer object requires skill and patience. First, a base is formed into a desired shape using either bamboo or a soft wood from a variety of trees. Secondly, a sap from a natural resin of the indigenous *Melanhorrea usitata* tree is collected. It is strained to obtain a suitable varnish, which is black in its natural state. A coat of lacquer is brushed on and allowed to dry until it is hardened. Then the piece is smoothed and polished. The lacquering process is repeated several times to achieve a suitable black surface for decoration. Although Burma and Thailand use a similar method of making lacquer, the country of origin manifests itself in the decoration.

A typical lacquerware betel container is a cylindrical covered box. The cover fits tightly over the body and both are about the same size and shape. A cursory glance does not reveal the two parts. The interior is fitted with two trays—one with small containers for the various ingredients and a pair of cutters; the other one for tobacco leaves. Betel leaves are stored in the bottom of the box. This form is made in a variety of sizes ranging from 8 to 35 centimetres (Colour Plate 15).

The characteristic colour of lacquer betel boxes is red.

Cinnabar is the source of colour in Burma, whereas the hill tribes living along the border between the two countries use ochre, a natural pigment found in the soil. Colour Plate 16 shows an ochre-based red betel box from the Karen hill tribes. Decoration is added in colours contrasting with the red. Yellow, green, orange, and brown obtained from indigenous minerals and plants are typical.

Designs are drawn with a free hand and often fill the sides, even on the interior, and the cover. Patterns can be divided into three groups: geometric, naturalistic, and pictorial. A design is enclosed in a series of bands on the top and bottom. Sometimes these bands are left a solid red colour to highlight the design they enclose. Geometric designs are characteristically repetitive and consist of elements linked together to form chains such as honeycombs and overlapping circles. A typical naturalistic pattern is a stylistic betel vine represented by pairs of leaves (in black) randomly spaced against an intricate background of undulating lines. Pictorial designs draw inspiration from the art, religion, and folklore of the culture. The *Ramayana* (the great Hindu epic) and the *Jataka* tales (stories of the former lives of the Buddha), the theatre, and drama are continuous sources of influence. Mythical creatures, such as celestial beings and sacred geese, are popular as are the Burmese signs of the zodiac (Colour Plate 17) and animals representing the days of the week.

Decoration is achieved by incising, gilding, moulding, or inlaying. Sometimes two or more methods are combined on the one piece. To decorate by incising, a design is cut into the surface of the black lacquered receptacle with a sharp tool. Then the incised area is filled with a contrasting colour, usually red. After drying, the excess pigment is wiped off and the container polished to a high finish. Other colours are added in the same way. Finally, a resinous sealer, which makes the box impervious to water, is applied and the surface polished to define the background.

Ceremonial betel sets are decorated with gold leaf in combination with either black or red lacquer. The method of application differs in Burma and Thailand. It begins with a lacquered container, which is naturally black. In Burma, a design is outlined on the surface with a water-soluble material, then paint is applied to the background and the remaining design lacquered. Thin squares of gold leaf are pressed on to the entire surface. After drying, the receptacle is washed to remove the excess gold leaf and water-soluble material. In Thailand, a design is drawn on the surface and covered with lacquer. Then gold leaf is pressed into the design before the lacquer dries. Finally, the excess is washed away.

Relief designs produced by moulding create a textured surface in contrast to the smooth surface of the previous method. Ash or husk or a similar material is added to the lacquer to make it pliable. Strands of lacquer are applied to the surface of a container to create a raised border. Designs, usually floral, are incised on the surface, then filled with strands of material moulded in the same pattern. Finally, the entire object is lacquered to seal the design. Colour Plate 18, a Burmese ceremonial betel box, is a fine example of the relief-moulded technique.

Sometimes this type of betel box is made even more elaborate by adding chips of coloured mirror glass. These are applied in the cavities between the outlines and adhered with lacquer. Another coat of lacquer seals the glass. Then gold leaf is added, using the same procedure to wash it and remove the excess. Another variation of relief-moulding is the addition of silver figures which stand out strikingly against a red background (Colour Plate 19).

An unusual lacquer piece is an octagon-shaped tray. It has two tiers with portions of the wood cut out on two sides, creating a pedestal effect. Strips of ochre lacquer divide the design into eight registers on each tier. A stylized geometrical design is inlaid in the registers using small pieces of red and brown painted wood (Plate 14). Although the provenance of this tray is unknown, the intricate design and careful workmanship suggest it belongs to the rich lacquer tradition of Burma or northern Thailand.

'Mother-of-pearl' is another type of inlay used for decorating betel boxes made of both lacquer and wood. It is particularly popular in Thailand and Vietnam as the coastline of these countries provides a plentiful source of the shellfish containing pearl. Opaque white or pinkish mother-of-pearl chips are inserted on to the surface of a betel receptacle with a dark background. A design

14. Betel tray, inlaid wood and lacquer, Thailand. Diameter 16 centimetres. (Collection of Donald L. Petrie and Niyanee Srikanthimarak)

is drawn and transferred, in reverse, on to paper. The shell is cut into flat pieces and honed to bring out the colour before gluing it to a piece of wood for cutting the desired shape. Next, each piece of shell is put on tracing paper. A base coat of a clear sticky mixture is put on to the receptacle, made of either lacquer or wood. While it is still damp on the base material, the paper is pressed into the surface and smoothed. When it is dry, water is sprayed over the surface and the paper is peeled off. The ridge between the shell and the lacquer surface is filled with several layers of a paste made from pulverized charcoal and sap. After drying, the receptacle is polished. This process is repeated until the layer of shell is completely concealed. A final polishing gives a rich black background contrasting against a lustrous mother-of-pearl design.

This technique is used in Vietnam for making a rectangular- or square-footed betel tray of wood and inlaid with a mother-of-pearl design of flowers and birds amongst clouds. A fine example of this technique can be seen on the doors of the ordination hall at Wat Phra Jetupon, popularly known as Wat Po, Bangkok's

15. Detail of a door with mother-of-pearl inlay at Wat Jetupon, Bangkok.

oldest and largest temple. A detail from the door depicts a betel set with small containers and a leaf holder (Plate 15).

Clay. Although fired clay was not widely used for making betel utensils, three countries—Cambodia, Thailand, and Vietnam— each with a long and distinguished ceramic tradition, excelled in making glazed stoneware containers for lime. Surprisingly, each country seems to have made only one shape (with slight variations), and each of these three shapes is entirely different from the others.

The method of making objects from clay has remained essentially unchanged since ancient times. Impurities are first sifted out before the clay is mixed to achieve a uniform consistency and to remove the air bubbles. Then it is ready for shaping. The forms are thrown on a wheel anchored on a base with a means for pivoting. After centring a lump of clay on the rotating wheel, the mass of clay is opened by pressing the thumbs into it. The walls of the vessel are then raised using even pressure with both hands. Finally, the lip is formed, and the vessel removed from the wheel.

After drying, it is decorated, glazed, and fired.

Quantities of glazed pots with bird appendages have been found dating from the Angkorian period of the Khmer civilization in Thailand (ninth–thirteenth centuries). Almost all of them have traces of lime on the interior. It is finely ground and hardened into a white paste, sometimes with a pinkish tinge, and readily disintegrates to a powder when scraped. A typical shape is a globular pot with an opening at the mouth and the applied beak, tail, and eyes of a bird. The body is grainy, sandy, and buff-coloured. Brown-glazed bird-shaped pots are the most common, although green-glazed examples are also known (Plate 16).

The type of lime container made at Si Satchanalai (Sawankhalok) by Thai potters in the fifteenth and sixteenth centuries is clearly in imitation of a bronze form. The tapering conical shape is made in two parts of about equal height with a lotus-bud knob (Plate 17). A similar form was made by potters of the Sankampaeng and Paan kilns in northern Thailand.

The form made in Vietnam is thickly potted with a spherical

16. Bird-shaped lime-pot, Khmer, twelfth century. Diameter 6 centimetres.

17. Lime containers, Thailand, fifteenth–sixteenth century. *Left*: bronze; *middle*: white monochrome; *right*: underglaze black. Average height 7 centimetres.

body and a heavy strap handle, an opening on the shoulder, and a carved foot-ring. The body is fine-grained and a pale buff colour. The handle and shoulder are decorated with modelled floral or animal forms. It is characteristically covered with a transparent cream-coloured glaze; crazing is typical (Colour Plate 20). Some examples have greenish splashes spilling over the handle and shoulder. Traces of lime are visible on the interior. Vietnam also made some underglaze blue porcelain holders for betel leaves.

Metal. The mineral deposits in South-East Asia have been sufficient to support the development of a metalworking craft that achieved a high degree of skilled workmanship. Minerals needed for alloys are available throughout the region and gold deposits are found in the peninsula, the Philippines, and parts of Indonesia. These, then, supply the necessary materials for making betel utensils of brass, bronze, silver, gold, and iron. From these metals, a diversified group of trays, boxes, small containers, cutters, and spittoons form a broad repertoire of betel chewing implements.

Brass and Bronze. A common method used for making brass betel boxes is the *cire perdue* or lost-wax process. A wax model forming the core of a mould is the basis for the name. Sheets of rolled wax are placed around the mould and thin strips are used for the decorative pattern. A cut is made through the form to remove the mould. A clay mixture is applied over the wax model and reinforced with additional clay. After drying, the cast is heated, causing the wax from the mould to melt and drain. Then molten brass is poured into the mould, replacing the 'lost-wax'. After cooling, the outer covering is broken and the brass object removed. Polishing completes the process.

Trengganu, on the east coast of Malaysia, and Mindanao, in the southern Philippines, are metalworking centres renowned for quality brassware. In Malaysia, two different compositions of brass are used for betel implements. Finely worked boxes are made of white brass, a combination of nickel and a high proportion of zinc. Sturdier, yet equally pleasing, betel sets are made of yellow brass, an alloy of zinc and scrap brass. Bronze, an alloy of copper and tin, is sometimes used in conjunction with brass, depending on the availability of metals.

A typical Malay betel set is made of yellow brass. It is rectangular with four short legs and fitted with a removable tray divided into compartments. Small round brass boxes with covers hold the individual ingredients. Sometimes leaves are stored under the tray. A betel cutter, also of brass, completes the set. Since Islamic religious tradition forbids the use of figurative designs, decorative patterns on Malaysian betel sets are either naturalistic or geometric, or in the Arabic script. The decoration on this example is a cut out arabesque pattern (Plate 18).

Brass betel boxes from Indonesia and Brunei are more complex in shape and more intricate in decoration than those found on the Malay Peninsula. Stylized animal designs integrated with vegetal and geometric motifs are favoured. On the coast of Borneo, betel ingredients are carried in small, crescent-shaped brass boxes made portable by tying them to the waist with strings.

The Maranao of the Philippines are renowned for their artistic skills and fine metalwork. A typical set from this area is cast by the

18. Betel set, brass, Malaysia. Length 17 centimetres.

lost-wax process. It is rectangular with a hinged lid and handles and fitted with three or four individual compartments for the ingredients. Decoration consists of a repetitive pattern dominated by a graceful floral scroll covering the entire box. Designs on these sets draw inspiration from wood-carving and textiles, both specialities of the Maranao people. The unique feature of this type of box is the use of silver inlay for the design (Plate 19).

Brass is also used for making pestles for pounding the ingredients of the betel quid to make them palatable to toothless chewers.

Metal is the most common material for the spittoon which accompanies many betel sets (see Plate 2). As Nieuhof put it, 'nice people spit in pots'. A spittoon, wrote the Norwegian Bock in the nineteenth century, is a 'cylindrical vessel of terra cotta, bronze, silver, or gold into which one may expectorate reddened saliva when the chewer is on "the white deck of a river steamboat"'. Westerners gave the homely, albeit descriptive, name of 'spit-box' or 'spit-tub' to the spittoon. 'A tall urn-shaped spit-box of brass is either in the midst of the circle or passing from one to another, that each may free her mouth from surplus saliva,'

19. Betel box, brass with silver inlay, Philippines. Length 21 centimetres.

noted one observer. 'In the houses of the better-class people various Quispitoors or spit–tubs always stand ready whether made of silver, metal, porcelain, or only of clay wherein to spit when Pinangh–chewing since this liquid otherwise cannot easily be removed from the floors', said Heydt.

Iron. The most common materials used for the blades of betel cutters are iron or steel because they give the necessary strength for cutting the areca–nut. To crack the nut, it is placed between the two blades and the handles pressed together. The handles may be sheathed in silver, gold, brass, or bronze to enable decoration. Betel cutters of the region are decorated with fanciful and creative motifs, although rarely as ingeniously as those from India.

An iron betel cutter from Thailand dating from the late four-teenth to fifteenth century is the oldest known example. The juncture is in the form of a bird. Traces of bronze sheathing and silver inlay are visible (Plate 20).

An elaborate Burmese betel cutter made of steel is decorated with inlaid stripes and stars of silver. A square of cross-hatching on each handle balances the design. Each pin is the centre of a

20. Betel cutter, iron, Thailand. Length 137 centimetres. (From the Samuel Eilenberg Collection; courtesy of the owner and his publisher, Hansjörg Mayer)

finely modelled crested bird (three of silver and one of steel). A Burmese inscription on the underside of the arm translates: 'master of a craft or other accomplishment' (Colour Plate 21).

A betel cutter from the island of Madura is made of iron and shaped like a crested bird with an elongated beak (Plate 21).

Silver and Gold. Deposits of silver and gold in Burma, Laos, Cambodia, Vietnam, Thailand, Malaysia, and parts of Indonesia provide the resources for a widespread art in betel sets of these metals. Recycled coins are another source of metals.

After a dinner hosted by the Governor of Siem Reap in 1871, 'there was music and dancing and then the Governor exhibited

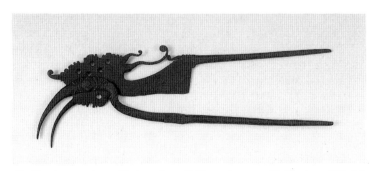

21. Betel cutter, iron, Madura. Length 28 centimetres. (Collection of Mohd. Yunus Noor)

his gold betel-boxes and other paraphernalia. He told us that all were made by a Cambodian goldsmith in Siamrap from the gold and silver coins of Hue, the capital of Annam, of entirely pure metal and they certainly were very elegant in design and finish. Articles included a large dish which contained the "kit", a gold betel-leaf and lime holder; a small gold tobacco-box; a silver cup with cigars; and a little silver box, made in the form of a fish, containing a perfumed ointment, used by the native noblemen to anoint their lips and nostrils,' wrote a European diplomat. A similar salve was popular with the upper-class Thais at the turn of the century. It resembles beeswax and is used to rub on the lips to prevent stinging from the areca-nut. This custom gave birth to a variety of miniature containers, finely worked and carefully decorated, such as the one in Plate 22 of silver and black coral.

The decorative techniques give each silver-producing area a ubiquitous style. Silversmiths from Padang and Aceh in Indonesia, for example, excel in filigree and those from Palembang in embossing, whereas Malaysian silversmiths produce repoussé with consummate skill. The artisans of Burma, Cambodia, Laos, and Thailand are experts in relief work. Whatever decorating method is used, it is usually worked on an alloy of silver and copper to increase its durability. The copper content causes silver objects to

22. Lip salve container, silver and black coral, Thailand. Diameter 2 centimetres. (Collection of Yadasji Wongpaiboon)

tarnish after prolonged exposure to the atmosphere. Several methods of decorating silver and gold in South-East Asia are employed.

A **chased** design is punched or beaten inside an outline with a flat-faced tool. This method depresses the area and produces a design in low relief. A typical background comprises clusters of small circles known as ring matting executed by punching. Chasing is commonly used in combination with other methods. An **embossed** design is hammered on the reverse side, then the background on the other side is filled to enhance the design. Embossed silver is often blackened with soot and oil to give definition. Bright colours made from metallic oxides are fused to a silver or gold base for an **enamelled** effect. This method is used to highlight features of a design. A line decoration cut into the surface of an object with a sharp tool is an **engraved** design. It is often used in combination with other techniques, particularly chasing and repoussé. Threads of silver wire are twisted into the desired shapes and then soldered together to produce an airy, open pattern known as **filigree**. To produce an **inlaid** design small pieces of a given material are inserted in a groove and burnished into position, giving an elaborate pattern. **Niello** ('black') is a method of decoration used on metal and distinguished by an intricate design which creates a light and dark effect. A black-coloured alloy of lead, silver, and copper is applied to a hammered design, then heat is used to fuse the compound to the silver or gold. The relief portion retains its original colour and the recessed area turns black. In an **open-work** design the background is beaten with a chisel allowing the surface motif to stand out. This action results in the pattern appearing in silhouette. A **repoussé** design is made by hammering the reverse side of sheet metal to produce a pattern in raised relief on the opposite side.

Metalworking in the Region

The art of metalworking in South-East Asia is an ancient one dating from the Dongson culture in Vietnam in the first millennium BC. The techniques of making tools from iron and bronze evolved to a sophisticated level of producing objects using various metals and methods of decoration. While the individual countries share similar materials and methods of working with metals, each one has achieved recognition for producing specific types of betel utensils.

In Burma, silver is the main medium for betel utensils besides lacquer. Burmese silversmiths are skilled craftsmen of relief techniques. Their intricate workmanship creates a depth of design unequalled in other silver making traditions. Lime boxes with covers are made in round, oval, hexagonal, and octagonal shapes and decorated with figures in high relief (Plate 23). Silver betel sets of the Shans, living in the hills of eastern Burma, normally bear a signature and a date. The style is less flamboyant and in lower relief than other Burmese silver work. A typical box is cylindrical with flaring bands at the top and bottom. It comprises three sections but, like the lacquer boxes, these are not clearly visible as the cover and body appear as one. The interior, fitted in a similar manner to the lacquer box, has a tray that holds small

23. Lime container, silver, Burma. (Courtesy Sylvia Fraser-Lu)

boxes for the individual ingredients; betel leaves are kept in the bottom of the box. Bold figures of animals and birds closely spaced around the sides dominate the design. Small floral and geometric motifs fill the remaining spaces. The top of the cover is equally embellished with raised figures of fine quality. A ceremonial betel box exquisitely worked in silver encapsulates the skills of Burmese silversmiths. Raised figures against a finely detailed background stand majestically on a base with an open-work design (Colour Plate 22).

Cambodian silversmiths excel in making small animal-shaped silver boxes for betel chewing ingredients. Characterized by delicate designs and careful workmanship, these boxes reflect the greatness in artistic skills belonging to the silversmiths' predecessors, the Khmers. A typical box is in the form of an animal in a crouched position. It is made in two parts with the cover forming the upper half of the animal. The join is barely visible and the box gives the impression of being one piece. Features of the animal are carefully worked using several techniques to give variety, depth, and texture. The average length is 10 centimetres. Popular shapes are mythical animals, a kneeling elephant with an upturned trunk, a reclining deer, a chicken, and a turtle (Plate 24). Fruit and vegetable forms, such as the mangosteen or pumpkin, are also made. A unique Cambodian lime container in silver is gourd-shaped and decorated with panels of flowers. It has an inscription and a shopmark on the base. The proportions, workmanship, and designs in high relief are of exceptionally fine quality. A lining of tin is added to avoid corrosion of the lime when it comes into contact with the silver (Plate 25).

Nestled between four other countries each with a full-blown silver tradition, an interchange of influences on Laos is inevitable. Laotian silver betel boxes are decorated by engraving, chasing, repoussé, or embossing, and sometimes gilding or enamelling with gold, blue, or green. Open-work with perforated motifs is typically Laotian. The body of the container is characteristically divided into registers and filled with a design. Motifs, either abstract or naturalistic, derive from Buddhist iconography. They include zodiac and mythical animals, flowers (especially the lotus),

24. Betel boxes, animal-shaped, silver, Khmer. (Photograph Widhaya Chaicharnthipayudh)

25. Lime container with liner, silver, Cambodia. Height 10 centimetres.
(Collection of Donald L. Petrie and Niyanee Srikanthimarak)

birds, the flame design, and curls resembling the hair found on
models of religious figures. Two shapes of betel receptacles distin-
guish Laotian craftsmanship. First, a tall, tubular container for lime
uses buffalo horn as a base with intricate silver work on the cover,
at the juncture, and around the base. The domed cover is shaped
like a lotus bud and is about one-third of the total height. It is
connected to the body by a chain. A silver spatula often accompanies this type of container. A ring on the tip allows the
lime container to be attached to a belt thus making it portable
(Colour Plate 23). Secondly, a holder for betel leaves in the shape
of a horn and decorated with repousse designs employing flowers,
clouds, and geometric motifs is found only in Laos (Plate 26).

Silver and gold have been associated with Indonesian culture
since the kingdoms of Srivijaya and Majapahit. During the period
of the Dutch East India trade in Indonesia, silver was fashioned
according to European taste. The two covered containers in
Plate 27, finely decorated with floral swags, reflect this influence.
Indonesian artisans excel in the silver techniques of repoussé and

26. Betel leaf holder, silver, Laos. Length 7 centimetres. (Collection of Donald L. Petrie and Niyanee Srikanthimarak)

27. Two betel boxes, silver, Indonesia, eighteenth century. (Collection CNO, Foundation for the cultural history of the Netherlands Overseas (CNO), c/o Rijksmuseum Amsterdam)

filigree. A typical betel set is faceted with six or eight sides set on a short pedestal. Small, round silver boxes with covers fill the interior. Betel leaves are stored in a triangular-shaped container. As much of the country is Muslim, decoration derives from nature and intermixes with geometric motifs. In other parts of Indonesia, the *wayang* or leather shadow puppet inspires the decoration.

Originally called 'Suvarnabhumi' ('the Golden Island or Peninsula'), the Malay Peninsula has a long and illustrious history of metalworking. Besides Malay craftsmen who produced betel sets decorated with traditional Islamic designs, immigrant Chinese silversmiths contributed to the output by creating an interesting blend of decorative motifs from both countries. These pieces represent a transitional period in the assimilation of the two cultures. Repoussé, open-work, and filigree are skilfully executed. Chinese craftsmen use mostly Buddhist motifs—the lotus, flame, and snail-like curls that look like the hair of the Buddha. They are also drawn towards traditional Chinese designs such as the immortals of the Daoist philosophy, dragons, and phoenixes. Malay craftsmen, on the other hand, mostly incorporated floral or vegetal motifs into their designs. A typical silver betel set is rectangular with the interior divided into compartments for individual containers and a drawer in the base for betel leaves. Small, round, covered boxes and cups are used for the individual ingredients. Plate 28 is a lime container crafted in Malay silver. The cylindrical form has a pinched waist and is decorated with a 'palmette' design; a chain connects the two parts of the box.

Artefacts recovered from ruins of the ancient capital of Ayutthaya confirm that the technique of working gold was at a sophisticated stage in the sixteenth century. Since then, gold betel sets have been part of royal regalia. Most silver work is done in the area of Chiang Mai in the north. Silver receptacles, small containers, and leaf holders for household use are finely made.

An exceptional small container is in the shape of Hanuman, the monkey-faced creature from the *Ramayana* (Plate 29). A local legend is associated with this figure. To decide where the capital city of Ayutthaya would be established, a dictum stated that an

28. Lime box, silver, from H. Ling Roth, *Oriental Silverwork: Malay and Chinese*, London, Truslove & Hanson, Ltd., 1910. (Courtesy Antiques of the Orient, Singapore)

29. Betel box, figure of Hanuman, silver, Cambodia. Diameter 7 centimetres. (Collection of Donald L. Petrie and Niyanee Srikanthimarak)

arrow should be shot from a bow. Hanuman caught it and pushed it into the ground. After this feat, his tail grew to an extraordinary length and the sweep of it formed the limits of the city. Even today, a 'Swinging Festival' is held annually to honour Hanuman.

A container for betel leaves, similar to those from Laos and Malaysia, is made in Thailand. It is in the shape of a truncated triangle or flattened cylinder and elaborately decorated, often with delicate filigree work. Themes derive from nature—floral scrolls, birds, flowers, grapes, and leaves (Colour Plate 24).

The technique of niello was also used to produce spectacular betel sets. Nakhon Si Thammarat, in southern Thailand, is the recognized centre of the craft. The art is practised to a lesser degree in the Malaysian states of Perak and Kedah and in Sumatra in Indonesia. Silver is the most common metal used for niello; gold is a recent alternative. Sometimes, these two metals are combined on a single piece. Nielloware is more ornate than ware made with only one metal. A fine example is a betel set in the collection of the late Prince and Princess Chumbhot of Nagara Svarga. It consists of a round tray decorated with a band of lotus petals and supported on four curved legs that culminate in the face of a mythical beast. Another base, under the legs, has a co-ordinating band of overlapping lotus petals. The fittings include a mortar and pestle, a knife, and four small pedestal containers (Colour Plate 25).

A betel set of gold, the most luxurious metal of the region, expresses the finest in Thai workmanship and materials. It is a delicate circular tray supported by four short round legs. It contains four small boxes divided into lobes and intricately decorated and a triangular-shaped leaf holder (Colour Plate 26).

Conclusion

THE lack of research on betel chewing in the twentieth century limits an assessment of its present-day use. Many people believe the custom is decreasing in urban areas of South-East Asia because of Western influence on lifestyles. In other parts of Asia, such as Bombay, the number of people who chew betel is actually increasing. A study of the domestic habits of Malaysians between 1940 and 1964 concluded that besides rice, betel was the most important item for daily use of a family living in rural areas. Today, though, betel chewing has declined and its importance in the family economy has diminished. There is, moreover, evidence that in areas with a strong missionary presence, such as parts of the Philippines and Papua New Guinea, betel chewing has declined in popularity. In Thailand, the custom diminished when the government passed a law in 1945 banning betel chewing and declared it an 'un-European and uncivilized' custom.

Discernible changes in the marketing of betel chewing ingredients reflect a response to changes in consumption. Five years ago, one did not have to go far in an urban area of South-East Asia to obtain the essential ingredients for a betel quid. A vendor on a street corner in the city selling leaves, nuts, and lime from a plastic bucket was a familiar sight. Today, the corner vendor is gone, indicating a decrease in the demand for the ingredients and, by deduction, a decrease in the custom.

There is also a discernible change in the ages of people who chew betel. In interviews with more than a hundred Asians between the ages of twenty and sixty, the most frequent remark was, 'I remember my grandmother chewing betel.' In Thailand, the custom seems to have decreased markedly with the past generation. The younger Thais, many of whom have been educated abroad and have inculcated Western ideas, find betel chewing no longer socially acceptable. They consider the custom old-fashioned and would be embarrassed if any Western friends saw their relatives chewing betel. Other modern social taboos, such as spitting, have contributed to the decline of betel chewing. Progress in urban areas has created an increased pace of life and discourages a leisurely chew.

Other changes in the custom of betel chewing are apparent in the twentieth century. The fast-food trend has extended to betel. In the past, a quid was always prepared in the home. Today, ready-made quids are preferred in many areas and their availability in local markets is increasing. One can buy machine-sliced nuts, spiced betel in airtight tins, and freeze-dried ingredients. These processed foods encourage the use of preserved, rather than fresh, products.

The impact of cigarette smoking on betel chewing remains controversial. Articles in journals often report that cigarette smoking has replaced betel chewing but fail to cite the basis for the statement. Others maintain that the introduction of tobacco has had little effect on betel chewing. One of the few surveys conducted on this aspect and reported by Anthony Reid in 1985 concluded that cigarette smoking has largely replaced betel chewing amongst adult Indonesian men. Women, though, according to the survey, continue to chew betel. Over 85 per cent of the men in Indonesia smoke cigarettes compared with 1.5 per cent of the women.

On the whole, evidence suggests that progress in the twentieth century may be eclipsing the custom of betel chewing. It is less universally practised in South-East Asia today and is decreasing in urban areas, and, to a lesser extent, in rural areas; it has declined more amongst men than women; and many of the present generation have never chewed betel. Despite these trends pointing towards a decline in the custom, the legacy of betel chewing remains and its use for medicinal and symbolical purposes continues.

Appendix

Regional Names for Betel Chewing Terms

Country	Betel Leaf	Areca-nut	Lime
Burma	kvám; kún	kùn si	thón
Cambodia	mlû	slâh	kombaur
China	laoye	binlang	shi–hui
India	paan	supari	chunam
Indonesia	sury	jambi	kapur
Laos	phû	mak	pûn
Malaysia	sireh, siri	pinang	kapur
Thailand	plû	mak	pûn
Vietnam	giâu	cau	vôi

Select Bibliography

Anand, Mulk Raj and Hutheesing, Krishna Nehru, *The Book of Indian Beauty*, Rutland, Vermont and Tokyo, Japan, Charles E. Tuttle, 1981.

Bellwood, Peter, *Man's Conquest of the Pacific*, New York, Oxford University Press, 1979.

Bickmore, Albert S., *Travels in the East Indian Archipelago*, New York, D. Appleton and Company, 1869; reprinted Singapore, Oxford University Press, 1991.

Bird, Isabella L., *The Golden Chersonese and the Way Thither*, London, John Murray, 1883; reprinted Kuala Lumpur, Oxford University Press, 1967, and Singapore, Oxford University Press, 1990.

Bock, Carl, *Temples and Elephants: The Narrative of a Journey of Exploration Through Upper Siam and Lao*, London, Sampson Low, Marston, Searle, & Rivington, 1884; reprinted Singapore, Oxford University Press, 1986.

Brownrigg, Henry, *Betel Cutters from the Samuel Eilenberg Collection*, Stuttgart and London, Edition Hansjörg Mayer, 1991.

Brus, R., 'The Royal Regalia of Thailand', *Arts of Asia*, September–October 1985, pp. 92–9.

Buddle, Anne, *Cutting Betel in Style* (exhibition catalogue), London, Victoria & Albert Museum, 1979.

Burkill, I. H., *A Dictionary of the Economic Products of the Malay Peninsula*, 2 vols., London, Crown Agents for the Colonies, 1935.

Chao (Ju-Kua), *Chau Ju-Kua: His Work of the Chinese and Arab Trade in the Twelfth and Thirteenth Centuries*, translated by F. Hirth and W. W. Rockhill, St. Petersburg, Printing Office of the Imperial Academy of Sciences, 1911.

Chia, Felix, *The Babas*, Singapore, Times Books International, 1980.

Chou Ta-Kuan (Zhou Daguan), *Notes on the Customs of Cambodia*, translated from the French version of Paul Pelliot by J. Gilman d'Arcy Paul, Bangkok, Social Science Association Press, 1967.

Cort, Mary Lovina, *The Heart of Farther India*, New York, Anson D. F. Randolf, 1886.

Curzon, George N., 'Journeys in French Indo-China (Tongking,

Annam, Cochin China, Cambodia)', Pt. 2, *The Geographical Journal*, Vol. II, No. 3, September 1893, pp. 193–218.

Dampier, William, *A New Voyage round the World, 1697*, edited by Sir Albert Gray, London, Argonaut Press, 1927.

De Bry, Theodore, *Petits Voyages* (Latin Edition), Pt. II, Frankfurt am Main, 1601.

De Haan, E., *Oud Batavai*, 2 vols., Batavia, G. Kolff, 1922.

De la Loubère, Simon, *A New Historical Relation of the Kingdom of Siam*, London, 1693; reprinted Kuala Lumpur, Oxford University Press, 1969, and Singapore, Oxford University Press, 1986.

Endicott, Kirk M., *An Analysis of Malay Magic*, Kuala Lumpur, Oxford University Press, 1970; reprinted Singapore, Oxford University Press, 1981.

Firth, Rosemary, *Housekeeping Among Malay Peasants*, 2nd edn., London, University of London and The Athlone Press, and New York, Humanities Press, 1966.

Fraser-Lu, Sylvia, *Burmese Lacquerware*, Bangkok, The Tamarind Press, 1985.

————, *Silverware of South-East Asia*, Singapore, Oxford University Press, 1989.

Galvao, Antonio, *The Discoveries of the World*, London, Hakluyt Society, 1862.

Gervaise, Nicolas, *The Natural and Political History of the Kingdom of Siam* (1688), translated and edited by John Villiers, Bangkok, White Lotus, 1989.

Gimlette, John D., *Malay Poisons and Charm Cures*, London, J. & A. Churchill, 1929; reprinted Kuala Lumpur, Oxford University Press, 1971.

Gorman, Chester F., 'Excavations at Spirit Cave, North Thailand, Some Interim Interpretations,' *Asian Perspectives*, Vol. 13, 1970, pp. 79–107.

Hamilton, Capt. Alexander, *A New Account of the East Indies, Being the Observations and Remarks of Capt. Alexander Hamilton*, 2 vols., Edinburgh, John Mofman, 1727.

Heydt, Johann Wolffgang, *Allerneuster Geographisch und Topographischer Schau Platz*, Nuremburg, 1744.

The Inscription of Ramkamhaeng the Great, edited by Chulalongkorn University on the 700th Anniversary of the Thai Alphabet, Bangkok, n.d.

Jain, Jyotindra, *Utensils: An Introduction to the Utensils Museum Ahmedabad*,

Ahmedabad, Surendra C. Patel for the Vechaar Foundation, 1984.

Khin Myo Chit, 'Betel Lore', *Sawaddi*, March–April 1979, pp. 12–14.

Klebert, Beowulf K., 'The Lerche Collection: Chewing Betel Through the Ages', *Arts of Asia*, January–February 1983, pp. 107–13.

Latham, Ronald (trans.), *The Travels of Marco Polo*, Penguin Books, London, 1958.

Lewin, L., *Phantastica*, London, Routledge & Kegan Paul, 1964.

Ling Roth, H., *Oriental Silverwork: Malay and Chinese*, London, Truslove & Hanson, Ltd., 1910; reprinted Kuala Lumpur, University of Malaya Press, 1966.

McCarthy, James, *Surveying and Exploring in Siam*, London, John Murray, 1902.

McNair, Major F., *Perak and the Malays*, London, Tinsley Brothers, 1878.

Ma Huan, *Ying-Yai Sheng-Lan (The Overall Survey of the Ocean's Shores) (1433)*, translated by J. V. G. Mills, Cambridge, Cambridge University Press, Hakluyt Society, 1970.

'Manasollasa', *Gaekwad Oriental Series*, Pt. 2, No. 14, edited by G. K. Shrigondekar, 1939.

Milner, G. (ed.), *Natural Symbols in South East Asia*, London, School of Oriental and African Studies, University of London, 1978.

Morarjee, Sumati, *Tambula: Tradition and Art*, Bombay, Tata Press, n.d.

Mouhot, Henri, 'Notes on Cambodia, the Lao country etc.', translated by Dr Thomas Hodgkin, *Journal of the Royal Geographical Society*, No. 32, 1862, pp. 142–63.

————, *Travels in the Central Parts of Indo-China (Siam), Cambodia, and Laos, During the Years 1858, 1859, and 1860*, 2 vols., London, John Murray, 1864; reprinted, Bangkok, White Lotus, 1986.

Neale, Frederick A., *Narrative of a Residence in Siam*, London, Office of the National Illustration Library, 1852; reprinted, Bangkok, White Lotus, n.d.

Nieuhof, Johan, *Voyages and Travels to the East Indies 1653–1670*, Singapore, Oxford University Press, 1988 (reprinted from the second part of Awnsham Churchill's, *A Collection of Voyages and Travels …*, London, 1704).

Penzer, N. M., *Poison-damsels, and Other Essays in Folklore and Anthropology*, London, private print for C. J. Sawyer, 1952.

Pigafetta, Antonio, *First Voyage around the World, 1524*, translated by J. A. Robertson, Manila, Filipiniana Book Guild, 1969.

The Proverbial Wisdom of Thailand, Vancouver, Asia Pacific Foundation of Canada, 1986.

Pym, Christopher, *The Ancient Civilization of Angkor*, New York, The New American Library, 1968.

Rathborne, Ambrose R., *Camping and Tramping in Malaya: Fifteen Years' Pioneering in the Native States of the Malay Peninsula*, London, Swan Sonnenschein, 1898; reprinted, Singapore, Oxford University Press, 1984.

Reid, Anthony, 'From Betel-Chewing to Tobacco-Smoking in Indonesia', *The Journal of Asian Studies*, Vol. XLIV, No. 3, May 1985, pp. 529–47.

Rooney, Dawn, *Khmer Ceramics*, Singapore, Oxford University Press, 1984.

Rubenstein, Carol, *The Honey Tree Song: Poems and Chants of Sarawak Dyaks*, Athens, Ohio University Press, 1985.

Soebiantoro, A. and Ratnatunga, M., *Folk Tales of Indonesia*, New Delhi, Sterling, 1977.

Sommerville, Maxwell, *Siam on the Meinam from the Gulf to Ayuthia*, London, 1897.

Teeuw, A. and Wyatt, D. K., *The Story of Patani*, 2 vols., The Hague, Martinus Nijhoff, 1970.

Theodoratus, Robert James, 'Betel Chewing', MA thesis, University of Washington, 1953.

Thierry, Solange, *Le Betel, I. Inde et Asie du Sud-Est*, Series K, Asie I, Paris, Musée de l'Homme, 1969.

Thompson, P. A., *Siam: An Account of the Country and the People*, 1910; reprinted, Bangkok, White Orchid Press, 1987.

Van Linschoten, John Huyghen, *The Voyage of John Huyghen van Linschoten to the East Indies from the Old English Translation of 1598*, Vol. 1 (No. 70), edited by Arthur Coke Burnell; Vol. 2 (No. 71), edited by P. A. Tiel, London, Hakluyt Society, 1885.

Vimol Bhongbhibhat (ed.), *The Eagle and the Elephant*, Bangkok, United Production, 1982.

Vincent, Frank, *The Land of the White Elephant: Sights and Scenes in South-East Asia 1871–1872*, New York, Harper and Brothers, 1874; reprinted Singapore, Oxford University Press in association with The Siam Society, Bangkok, 1988.

White, Joyce C., *Discovery of a Lost Bronze Age: Ban Chiang*, Philadelphia, The University Museum, University of Pennsylvania, and the Smithsonian Institution, 1982.

Windstedt, Richard, *The Malay Magician: Being Shaman, Saiva and Sufi*, London, Routledge and Kegan Paul, 1961; reprinted Kuala Lumpur, Oxford University Press, 1982.

Wong, Grace, 'A Comment on the Tributary Trade between China and Southeast Asia, and the Place of Porcelain in this Trade, During the Period of the Song Dynasty in China', *Chinese Celadons and Other Related Wares in Southeast Asia*, Singapore, Southeast Asian Ceramic Society, 1979, pp. 73–100.

Index

References in brackets refer to Plate numbers; those in brackets and italics to Colour Plate numbers.

Alexander, William, 17
Angkor, 5, 8
Areca catechu, 3, 13, 16, 17, 19, 23, 26, (1), (6), (7), (2), (3)
Ayutthaya, 3, 10, 63

Ban Chiang, 14
Bangkok, 3, 19, 25, 27, 48
Bantam, 22
Batak, 35, 43
Batavia, 10, 40
Betel set, (2), (5), (11), (15), (18), (10), (14), (25), (26); betel cutter, 52, 54, 55, (20), (21), (21); betel leaf holder, (26), (24); lime container, 22, 39, 41, 43, 44, 50, 59, 61, 63, (16), (17), (23), (25), (28), (11), (20), (23); mortar and pestle, 22, 40, 65, (9); spatula, 40, 61, (8); spittoon, 8, 40, 53; tray, 3, 12, 33, 38–40, 47, 48, 52, 58, 65, (14)
Betel slaves, 8, (3)
Betel vendor, (10), (5)
Bickmore, Albert, 16
Bird, Isabella, 1, 32, 50, 54, 55
Bloody Mary, 28
Bock, Carl, 11, 53
Borneo, 17, 23, 26, 38, 41, 52
Burkill, I. H., 13
Burma, 7, 8, 11, 12, 17, 21, 24, 34, 36, 40, 41, 45–7, 55, 56, 58

Cambodia, 15, 27, 33, 34, 38, 39, 49, 54–6

Cancer, 26, 27
Champa, 14, 24
Chemical properties: alkaloid, 25, 26; arecoline, 25; tannin, 25, 26
China, 2, 14, 24
Chou Ta-Kuan (Zhou Daguan), 5
Cire perdue, 52
Cort, Mary, 5, 6, 24
Curzon, George N., 1, 6

Da Gama, Vasco, 14
Dampier, William, 19
Dayak, 33
De Bry, Theodore, 8
De Haan, E., 6
De la Loubère, Simon, 7, 17, 19
Dutch East India Company, 10, 40
Duyong Cave, 14

Fibre: bamboo, 24, 32, 41, 43–5; rattan, 43; reeds, 41
Flores, 41
Folklore, 7, 34, 46

Galvao, Antonio, 10
Gambier, 23, 24, 38, 39
Gervaise, Nicolas, 1, 3

Hamilton, Alexander, 10
Heydt, John, 19, 54

Ibn Batuta, 14
India, 5, 10, 11, 13, 14, 23, 24, 27, 28, 37, 40, 54, 61

Indonesia, 10, 13, 14, 22, 23, 41, 51, 52, 55, 56, 61, 63, 65, 67

I'Tsing, 23

Jataka stories, 46
Java, 17, 22, 24, 28, 36, 38, 45, 55
Johore, Sultan of, 8

Kelantan, 38
Khin Myo Chit, 6, 35

Lacquer, 40, 41, 44–8, 58
Laos, 22, 55, 56, 61, 65

Ma Huan, 5
McCarthy, James, 3
McNair, Major F., 10
Magellan, Ferdinand, 6
Mahabharata, 36
Malacca, 7, 11
Malaysia, 8, 10, 12–14, 19, 23, 24, 26, 31, 33–5, 38, 52, 55, 56, 65
Maldives, 34
Marco Polo, 7, 14
Medicine, 2, 5, 20, 26
Metal, 11, 41, 51, 53, 54, 56, 57, 65; brass, 39, 41, 51–4; bronze, 45, 50–4, 57; gold, 3, 8, 10, 12, 13, 22, 45–7, 51, 53–7, 59, 61, 63, 65; iron, 51, 54, 55, 57; silver, 3, 8, 10, 47, 51, 53–9, 61, 63, 65
Metalwork: chased, 57; embossed, 57; enamelled, 12, 57; engraved, 57; filigree, 56, 57, 63, 65; inlaid, 3, 12, 47, 53, 54, 57; niello, 57, 65; open-work, 57, 59; repoussé, 56, 57, 59, 61, 63
Mindanao, 43, 52
Morarjee, Sumati, 5, 6
Mother-of-pearl, 47, 48
Mouhot, Henri, 8

Natural forms: coconut, 21, 22, 32, 41; gourd, 59; horn, 41, 43, 61
Neale, Frederick, 25, 27

Nieuhof, Johan, 6, 17, 21, 53

Palembang, 56
Papua New Guinea, 2, 43, 66
Penang, 24
Penzer, N. M., 13
Perak, 10, 38, 65
Perlis, Raja of, 12
Philippines, the, 6, 14, 17, 19, 22, 33, 41, 43, 44, 51, 52, 54, 66
Pigafetta, Antonio, 6
Piper betle, 19, 26, (8), (9), (4)
Proverb, 22, 34, 35, 37

Quid: areca-nut, 30, 31, 33, 34, 48; betel leaf, 1, 6, 20, 21, 26, 28, 30, 34–7; lime, 1–3, 16, 22, 23, 28, 32, 34, 37–41, 43, 44, 49–51, 56, 58, 59, 61, 63, 66

Ramayana, 46, 63
Rathborne, Ambrose B., 23
Rites of Passage: birth, 26, 33, 56; betrothal, 35, 37; marriage, 35, 37–9; death, 33, 34, 41

Sarawak, 3, 39
Schouten, Justus, 10
Si Satchanalai, 50
Singapore, 10, 24
Sommerville, Maxwell, 3
Spirit Cave, 13
Sulawesi, 41
Sumatra, 3, 17, 23, 35, 38, 43, 65

Teeth, 6, 16, 23, 26–9; black teeth, 27, 28; gum disease, 28
Ternate, 10
Thailand, 3, 7, 10–14, 17, 19, 22–4, 30, 31, 33–5, 38–41, 44–7, 49, 50, 54–6, 65, 66
Thompson, P. A., 10
Tikopia, 2
Timor, 38, 41, 43
Tobacco, 23, 32, 45, 56, 67

Trengganu, Sultan of, 13, 52

Van Linschoten, John Huyghen, 5, 8,
   10, 19, 27, 37
Vietnam, 14, 15, 22, 37, 47–51, 55,
   57
Vincent, Frank, 3

Wood, 7, 19, 32, 41, 43–5, 47, 48, 53

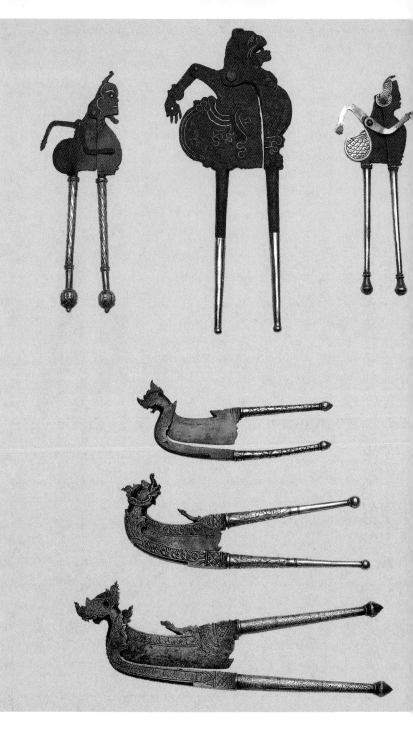